P9-CFO-718

Light at the Edge of the World

Morning light on the north face of Makalu,
with silver firs in shadow, Kama Valley, Tibet, 2000

LIGHT AT THE EDGE *of* THE WORLD

A JOURNEY THROUGH THE REALM OF VANISHING CULTURES

Text and photographs by

WADE DAVIS

NATIONAL GEOGRAPHIC
WASHINGTON, D.C.

CHICAGO HEIGHTS PUBLIC LIBRARY

306.08
D26
c.1

A YOUNG BOY, CHINCHERO, PERU, 1982

CONTENTS

5-28-02 B&T 35.00

THIS BOOK RECOUNTS, visually and in words, journeys made over the course of my work as an ethnobotanist and anthropologist, student and writer. The initial chapters, covering travels through northern Canada, the high Andes, Amazon, Orinoco, and Haiti, touch upon those seminal experiences and encounters that led me as a student to appreciate and embrace the key revelation of anthropology, the idea that distinct cultures represent unique visions of life itself, morally inspired and inherently right. The latter half of the book, those sections dealing with the plight of the Penan in Borneo, the pastoral nomads of northern Kenya, and the fate of Tibet, suggests something of the dark undercurrent of our age, the manner in which ancient peoples throughout the world are being torn from their past and propelled into an uncertain future. The book ends with the redemptive promise of Nunavut, the Inuit homeland recently established in the Canadian Arctic.

Although this is primarily a book of photographs, I should stress from the outset that these images were taken over a span of twenty-five years, and at no time was photography my principal pursuit. An ethnographer seeks to distill the essence of a people, a poetic sense of the wonder of who they are, the unique qualities that allow them to live, the pressures and challenges that urge them forward. A photographer writes with light, paying attention to a passing shadow, the incipient shift of forms within the frame, that flash of intuition and spontaneity that allows the image to become the master of the moment. Although both vocations imply, as Susan Sontag might suggest, aggressive acts of non-intervention, they are, for the most part, mutually exclusive endeavors. Great photographs, wrote Henri Cartier-Bresson, come about in that fraction of a second when the head, heart, and eye find perfect alignment in an axis of the spirit. The anthropological lens rarely achieves such precise focus.

In presenting these images, most of them published here for the first time, I have also gathered together those stories, drawn from years of travel, that complement them and best illustrate the central theme. Certain passages, those describing the lives of the Kogi and Ika in Colombia and the sense of spirit and landscape that resonates in the Sierra Nevada de Santa Marta and along the spine of the Andes, echo sentiments from my earlier books, presented here in a different form; the account of the Penan, Ariaal, and Rendille first appeared in the pages of *National Geographic* magazine. Richard Evans Schultes will be familiar to readers of *One River*. Vodoun has been the subject of two previous books and several essays, but nothing that I have written captures the power of the faith, indeed its ethereal beauty, as well as the images of the devout acolytes, awash in the sacred waterfall of Saut d'Eau.

It is my hope that these photographs and stories will provide a moving and visceral sense of the key issue celebrated by the book, the wondrous diversity and character of the ethnosphere, a notion perhaps best defined as the sum total of all thoughts, beliefs, myths, and intuitions made manifest today by the myriad cultures of the world. The ethnosphere is humanity's greatest legacy. It is the product of our dreams, the embodiment of our hopes, the symbol of all that we are and all that we have created as a wildly inquisitive and astonishingly adaptive species.

My travels through the ethnosphere have, for the most part, been driven by simple curiosity. The mysterious formula of a folk preparation, and the thought that zombies might actually exist, first brought me to Haiti. The study of coca, a plant known to the Inca as the Divine Leaf of Immortality, took me the length of the Andean Cordillera, a journey of well over a year. A desire to understand something of the shamanic art of healing resulted in a sojourn of many months in the Northwest Amazon. A chance to photograph the clouded leopard, rarest of all the great cats, took me to the Himalayas and the Kangshung face of Everest. The Winikina-Warao, canoe people whose feet rarely touch dry land, drew me to the Orinoco Delta of Venezuela. A longing to stand in the light of the midnight sun at the mouth of the Northwest Passage and hunt narwhal with Inuit led to the high Arctic and the ice islands that haunt the memory of Europe.

In every case, the scientific quest served as a metaphor, a lens through which to interpret a culture and acquire personal experience of the other. But what ultimately inspired these journeys was a restless desire to move, what Baudelaire called "the great malady," horror of home. Simply put, I sought escape from a monochromatic world of monotony, in the hope that I might find in a polychromatic world of diversity the means to rediscover and celebrate the enchantment of being human.

At the Pang La Pass, looking north toward
Chomolungma, Tibet, 2000

The spirit of Iemanjá, Afro-Brazilian goddess of water, at a New Year's Eve ceremony on the Río Negro near Manaus, Brazil, 1982

CHAPTER ONE

THE WONDER OF THE ETHNOSPHERE

One night on a ridge in Borneo, close to dusk, with thunder over the valley and the forest alive with the electrifying roar of black cicadas, I sat by a fire with Asik Nyelit, headman of the Ubong River Penan, one of the last nomadic peoples of Southeast Asia. The rains, which had pounded the forest all afternoon, had stopped, and the light of a partial moon filtered through the branches of the canopy. Earlier in the day, Asik had killed a barking deer. Its head lay roasting in the coals.

At one point, Asik looked up from the fire, took notice of the moon, and quietly asked me if it was true that people had actually journeyed there, only to return with baskets full of dirt. If that was all they had found, why had they bothered to go? How long had it taken, and what kind of transport had they had?

It was difficult to explain to a man who kindled fire with flint—and whose total possessions amounted to a few ragged clothes, blowpipe and quiver of poisoned darts, rattan sleeping mat and carrying basket, knife, ax, parang blade, loincloth, plug of tobacco, three dogs, and two monkeys—a space program that had consumed the energy of nations and, at a cost of nearly a trillion dollars, placed twelve men on the moon. Or the fact that over the course of six missions, they had traveled a billion and half miles and, indeed, brought back nothing but rocks and lunar dust, 828 pounds altogether.

One small fragment of this precious cargo found its way to Washington, D.C., where it is today embedded in a swirl of blood-red crystal, the focal point of a beautiful stained-glass window of the National Cathedral, one of the largest and most dramatic Gothic churches in the world. When I first visited the cathedral and sat beneath its soaring vault, moved by the perfect harmony of its architecture and the shafts of light flooding the nave, I thought of Asik and his queries. The panel of glass known as the Space Window is dense with primary colors, circles of the deepest blues and reds representing the spheres of the heavens, with small crystals flaring on all sides. In the Gothic tradition, the light pouring through these windows is the Light Divine, a mystic revelation of the spirit of God. In contrast to this sacred luminosity, the tiny moon rock appears cold and lifeless, black, inert.

Here, perhaps, was an answer to Asik's question. The true purpose of the space journeys, or at least their most profound and lasting consequence, lay not in wealth secured but in a vision realized, a shift in perspective that would change our lives forever. The seminal moment occurred on Christmas Day, 1968, a full six months before the first lunar landing, as the crew of Apollo 8 emerged from the dark side of the moon to see rising over its surface a small and fragile blue planet, floating, as one astronaut would recall, in the velvet void of space. For the first time in history, our world was revealed: a single interactive sphere of life, a living organism composed of air, water, and soil. This transcendent vision, more than any amount of scientific data, taught us that the Earth is a finite place that can endure our neglect for only so long. Inspired by this new perspective, this new hope, we began to think in new ways, a profound shift in consciousness that in the end may well prove to be the salvation of a lonely planet.

Consider how far we have come. Forty years ago, the environmental movement was nascent. Highway beautification was a key initiative and just convincing motorists to stop throwing garbage out of car windows was considered a great victory. Writers such as Rachel Carson, who warned of far more dire scenarios, were lone voices in the wild. Gary Snyder, whose poetry touched that place of sensual memory reached later by the prophets of deep ecology, used to hitchhike across the United States simply to spend an evening with someone to whom he could relate. No one thought of the ozone layer, let alone of our capacity to destroy it and thus compromise the very conditions that make life possible. A mere decade ago, scientists who warned of the greenhouse effect were dismissed as radicals. Today, it is those who question the existence and significance of climate change who occupy the lunatic fringe. Twenty years ago, "biodiversity" and "biosphere" were exotic terms, familiar only to a small number of earth scientists and ecologists. Today, these are household words understood and appreciated by schoolchildren. The biodiversity crisis, marked by the extinction of over a million life forms in the past three decades alone, together with the associated loss of habitat, has emerged as one of the central issues of our times.

What stands out in this checkered history is not merely the pace of attitudinal change, but its dramatic scale and character. A single generation has witnessed a shift in perspective and awareness so fundamental that to look back is to recall a world of the blind. Evidence of the impending environmental crisis swirled all around us, but we took little notice. Fortunately, we have come to see, at least partially; and though solutions to the major environmental problems remain elusive, no nation or government can ignore or deny the magnitude of the threat or the urgency of the dilemma. This alone represents a reorientation of human priorities that is both historic in its significance and profoundly hopeful in its promise.

ACKNOWLEDGING A PROBLEM, of course, is not the same as finding a solution. And having one veil lifted from our vision does not necessarily mean that we have fully recovered our sight. Even as we lament the collapse of biological diversity, we pay too little heed to a parallel process of loss, the demise of cultural diversity, the erosion of what might be termed the ethnosphere, the full complexity and complement of human potential as brought into being by culture and adaptation since the dawn of consciousness. As linguist Michael Krauss reminds us, the most pessimistic biologist would not dare suggest that half of all extant species are endangered or on the edge of extinction. Yet this, the most apocalyptic assessment of the future of biological diversity, scarcely approaches what is known to be the best conceivable scenario for the fate of the world's languages and cultures.

Worldwide, some 300 million people, roughly five percent of the global population, still retain a strong identity as members of an indigenous culture, rooted in history and language, attached by myth and memory to a particular place on the planet. Though their populations are small, these cultures account for 60 percent of the world's languages and collectively represent over half of the intellectual legacy of humanity. Yet, increasingly, their voices are being silenced, their unique visions of life itself lost in a whirlwind of change and conflict.

There is no better measure of this crisis than the loss of languages. Throughout all of human history, something in the order of ten thousand languages have existed. Today, of the roughly six thousand still spoken, fully half are not being taught to children, meaning that, effectively, they are already dead, and only three hundred are spoken by more than a million people. Only six hundred languages are considered by linguists to be stable and secure. In another century, even this number may be dramatically reduced.

More than a cluster of words or a set of grammatical rules, a language is a flash of the human spirit, the filter through which the soul of each particular culture reaches into the material world. A language is as divine and mysterious as a living creature. The biological analogy is apropos. Extinction, when balanced by the birth of new species, is a normal phenomenon. But the current wave of species loss due to human activities is unprecedented. Languages, like species, have always evolved. Before Latin faded from the scene, it gave rise to a score of diverse but related languages. Today, by contrast, languages are being lost at such a rate, within a generation or two, that they have no chance to leave descendants. By the same token, cultures have come and gone through time, absorbed by other more powerful societies or eliminated altogether by violence and conquest, famines, or natural disasters. But the current wave of assimilation and acculturation, in which peoples all over the Earth are being drawn away from their past, has no precedent.

Of the more than two thousand languages in New Guinea, five hundred are each spoken by fewer than five hundred people. Of the 175 native languages still alive in the United States, 55 are spoken by fewer than ten individuals. The words and phrases of only twenty are whispered by mothers to their babies. Of the eighty languages in California at the time of European contact, fifty remain, but not one is today spoken by a child. In Canada, there were once some sixty indigenous languages, but only four remain viable: Cree, Ojibwa, Dakota, and Inuktitut. In all North America, only one native language, Navajo, is spoken by more than a hundred thousand individuals.

What could possibly be more lonely than to be enveloped in silence, to be the last person alive capable of speaking your native tongue, to have no means of communicating and no chance of telling the world of the wonders you once knew, the wisdom and knowledge that had been passed down through generations, distilled in the sounds and words of the elders? Such is the fate, in fact, of many people; for every two weeks somewhere in the world, a language is lost. Even as you read these words, you can hear the last echoes of Kasabe in Cameroon, Pomo in California, Quinault in Washington state, Gosiutes in Utah, Ona in Patagonia, and scores of other languages that in the West do not even have a name.

The vast majority of the world's languages have yet even to be chronicled. In Papua New Guinea, only a dozen of the eight hundred languages have been studied in detail. Worldwide, perhaps as many as four thousand languages remain inadequately described. The cost of properly doing so has been estimated by linguists at $800 million, roughly the price of a single Aegis Class navy destroyer. In the United States, journalists devote columns of print to the fate of the spotted owl, but scarcely a word to the plight of the world's languages. The United States government spends $1 million a year attempting to save a single species of wildlife, the Florida panther, but only $2 million a year for the protection of all the nation's indigenous tongues, from the Arctic slope of Alaska to the pine barrens of Florida, from the deserts of the Navajo to the spruce forests of Maine. And yet, each language is, in itself, an entire ecosystem of ideas and intuitions, a watershed of thought, an old-growth forest of the mind. Each is a window into a world, a monument to the culture that gave it birth, and whose spirit it expresses. When we sacrifice a language, notes Ken Hale, a professor of linguistics at the Massachusetts Institute of Technology, we might as well drop a bomb on the Louvre.

The ultimate tragedy is not that archaic societies are disappearing but rather that vibrant, dynamic, living cultures and languages are being forced out of existence. At risk is a vast archive of knowledge and expertise, a catalogue of the imagination, an oral and written literature composed of the memories of countless elders and healers, warriors, farmers, fishermen, midwives, poets, and saints. In short, the artistic, intellectual, and spiritual expression of the full complexity and diversity of the human experience.

To place a value on what is being lost is impossible. The ecological and botanical knowledge of traditional peoples, to cite but one example, has obvious importance. Less than one percent of the world's flora has been thoroughly studied by science. Much of the fauna remains unknown. Yet a people such as the Haunóo, forest dwellers from the island of Mindoro in the Philippines, recognize more than 450 animals and distinguish 1500 plants, 400 more than are recognized by Western botanists working in the same forests. In their gardens grow 430 different cultigens. From the wild, they harvest a thousand species. Their taxonomy is as complex as that of the modern botanist, and the precision with which they observe their natural environment is, if anything, more acute.

Such perspicacity is typical of indigenous peoples. Native memory and observation can also describe the long-term effects of ecological change, geological transformations, even the complex signs of imminent ecosystem collapse. Aborigine legends record that once it was possible to walk to the islands of the Coral Sea, to reach Tasmania by land, facts confirmed by what we now know about sea level fluctuations during the Ice Age. In the high Arctic, I once listened as a monolingual Inuit man lamented the shifts in climate that had caused the weather to become wilder and the sun hotter each year, so that for the first time the Inuit suffered from skin ailments caused, as he put it, by the sky. What he described were the symptoms and consequences of ozone depletion and global warming.

Elsewhere in Canada, in the homeland of the Micmac, trees are named for the sound the prevailing winds make as they blow through the branches in the fall, an hour after sunset during those weeks when the weather comes always from a certain direction. Through time, the names can change, as the sounds change as the tree itself grows or decays, taking on different forms. Thus, the nomenclature of a forest over the years becomes a marker of its ecological health and can be read as a measure of environmental trends. A stand of trees that bore one name a century ago may be known today by another, a transformation that may allow ecologists, for example, to measure the impact of acid rain on the hardwood forests.

Some botanists suggest that as many as forty thousand species of plants may have medicinal or nutritional properties, a potential that in many instances has already been realized by indigenous healers. When the Chinese denounced Tibetan medicine as feudal superstition, the number of practitioners of this ancient herb-based discipline shrank from many thousands to a mere five hundred. The cost to humanity is obvious. But how do you evaluate less concrete contributions? What is the worth of family bonds that mitigate poverty and insulate individuals from loneliness? What is the value of diverse intuitions about the cosmos, the realms of the spirit, the meaning and practice of faith? What is the economic measure of a ritual practice that results in the protection of a river or a forest?

Answers to these questions are elusive, impossible to quantify; and as a result, too few recognize the full significance and meaning of what is being lost. Even among those sympathetic to the plight of small indigenous societies, there is a mood of resignation, as if these cultures are fated to slip away, reduced by circumstance to the sidelines of history, removed from the inexorable progression of modern life.

Though flawed, such reasoning is perhaps to be expected, for we are all acolytes of our own realities, prisoners of our perceptions, so blindly loyal to the patterns and habits of our lives we forget that, like all human beings, we too are enveloped by the constraints and protection of culture. It is no accident that the names of so many indigenous societies—the Waorani in the forests of the Northwest Amazon, the Inuit of the Arctic, the Yanomami in the serpentine reaches of the upper Orinoco—translate simply as "the people," the implication being that all other humans by default are non-people, savages and cannibals dwelling at the outskirts of the known world. The word "barbarian" is derived from the Greek *barbarus*, meaning "one who babbles," and in the ancient world, it was applied to anyone who could not speak the language of the Greeks. Similarly, the Aztec considered all those incapable of understanding Nahuatl to be mute. Every culture is ethnocentric, fiercely loyal to its own interpretation of reality. Without such fidelity, the human imagination would run wild, and the consequences would be madness and anarchy.

But now, equipped with a fresh perspective, inspired in part by this lens brought to us from the far expanses of space, we are empowered to think in new ways, to reach beyond prosaic restraint and thus attain new insight. To dismiss indigenous peoples as trivial, to view their societies as marginal, is to ignore and deny the central revelation of anthropology.

In Haiti, a Vodoun priestess responds to the rhythm of drums and, taken by the spirit, handles burning embers with impunity. In the Amazonian lowlands, a Waorani hunter detects the scent of animal urine at forty paces and identifies the species that deposited it in the rain forest. In Mexico, a Mazatec farmer communicates in whistles, mimicking the intonation of his language to send complex messages across the broad valleys of his mountain homeland. It is a vocabulary based on the wind. In the deserts of northern Kenya, Rendille nomads draw blood from the faces of their camels and survive on a diet of milk and wild herbs gathered in the shade of frail acacia trees. In Borneo, children of the nomadic Penan watch for omens in the flight of crested hornbills. On an open escarpment in the high Arctic, Inuit elders fuse myth with landscape, interpreting the past in the shadow of clouds cast upon ice.

Just to know that such cultures exist is to remember that the human imagination is vast, fluid, infinite in its capacity for social and spiritual invention. Our way of life, with its stunning technological wizardry, its cities dense with intrigue, is but one alternative rooted in a particular intellectual lineage. The Polynesian seafarers who sense the presence of distant atolls in the echo of waves, the Naxi shaman of Yunnan who carve mystical tales into rock, the Juwasi Bushmen who for generations lived in open truce with the lions of the Kalahari, reveal that there are other options, other means of interpreting existence, other ways of being.

Every view of the world that fades away, every culture that disappears, diminishes a possibility of life and reduces the human repertoire of adaptive responses to the common problems that confront us all. Knowledge is lost, not only of the natural world but of realms of the spirit, intuitions about the meaning of the cosmos, insights into the very nature of existence.

Ipeelie Koonoo on an ice floe, scanning the sea for narwhal at Admiralty Inlet, Baffin Island, Nunavut, 1997

The eyes of Rara at the town of Desdunnes,
Artibonite Valley, Haiti, 1982

On a wedding day, Ariaal boys await the ritual slaughter of a bull, Mount Marsabit, Kenya, 1998

Machu Picchu, Peru, 1998

A packhorse at Cold Fish Lake,
Spatsizi wilderness, British Columbia, 1980

CHAPTER TWO

THE EYELIDS OF WOLVES

As a child, I was impressed mostly by the cold night air, the ice cracking on the river, and the lights of the Catholic seminary on the village point, where the stone windmill stood frozen by the winter. My entire world was limited to a dozen or so suburban blocks, a warren of brick houses and asphalt that sprawled over the remnants of old Québec, a rock quarry once worked by peasants, wheat fields and orchards where priests lingered on hot summer days, dirt trails that followed traplines and the paths of the *coureurs de bois*, the fur traders who broke open a continent. In my dreams, I wandered with them, up the St. Lawrence River to the Ottawa, past the islands of Georgian Bay to the Superior lakehead, and beyond to the far reaches of the Athabaska territory, through the lands of the Huron and Cree, Ojibway and Kaska. My waking hours were given over to more prosaic concerns: school and endless games of pickup baseball, football, and hockey, the seasonal pursuits that marked the annual round for the English community of Pointe Claire.

When, years later, I returned as an adult, what astonished me most was to realize how small my universe had been and how intimately I had known it. Every blade of grass resonated with a story. Shadows marked the ground where trees had fallen in my absence. Innovations and new construction I took as personal insults, violations of something sacred that lay at the confluence of landscape and memory. Never would I know a place so completely, embrace it with such intensity. Yet the thought of never having left, of having stayed behind as some of my old friends and neighbors had done, left me shuddering with dread. For at its core, Pointe Claire remained what it had been in my father's time, a bedroom community of harried commuters where the English did not speak to the French, and the French looked across a deep cultural divide to a society they despised. I say this not in judgment but merely to stress how narrow were the limits of my world.

At the age of eleven, I joined the commuters on the morning train, dressed like them in dark jacket and tie, heading into the city to the first of a series of respectable private schools that taught me too much of what I did not want to know and just enough of what I did. And that was to get away, the sooner the better. A first break occurred in the summer of 1968, when a Spanish teacher took six of us to Colombia. The teacher was English by birth, dapper in appearance, with a scent of cologne that in those days gave him the fey veneer of a dandy, an impression betrayed by the scars on his face and a glass eye that marked a body blown apart in the war. His name was John Forester.

At fourteen, I was the youngest of his group and the most fortunate, for unlike the others who spent a sweltering season in the streets of Cali, I was billeted with a family in the mountains above the plains, at the edge of trails that reached west to the Pacific. It was a typical Colombian scene: a flock of children too numerous to keep track of, an indulgent father half the size of his wife, a wizened old grandmother who muttered to herself on a porch overlooking fields of cane and coffee, a protective sister who more than once carried her brother and me home half drunk to a mother, kind beyond words, who stood by the garden gate, hands on hips, feigning anger as she tapped her foot on the stone steps. For eight weeks, I encountered the warmth and decency of a people charged with a strange intensity, a passion for life, and a quiet acceptance of the frailty of the human spirit. Several of the other Canadian students longed for home. I felt as if I had finally found it.

Each Sunday, there were dances and wild moments when horsemen from a dozen villages raced over parched fields and along dusty roads where women offered food and teased the riders with their beauty. Though school was out for the summer, one teacher convened classes in his house and discreetly introduced themes that could not be embraced in the open: the plight of the poor, the meaning of a phrase of poetry, the fate of Che Guevara, recently killed in Bolivia. And there were darker moments: the sight of beggars, limbs swollen with disease, and armed soldiers beating ragged children, feral as alley cats, as they scattered into a black night cracked by gunfire.

Life was real, visceral, dense with intoxicating possibilities. I learned that summer to have but one operative word in my vocabulary, and that was yes to any experience, any encounter, anything new. Colombia taught me that it was possible to fling oneself upon the benevolence of the world and emerge not only unscathed but transformed. It was a naive notion, but one that I carried with me for a long time.

~

SOME YEARS LATER, after finishing high school in British Columbia, where my father had been raised, the son of a doctor in a small mining town in the Canadian Rockies, I returned east to attend university at Harvard. When the time came to select an academic major, serendipity played a hand. Faced with a dazzling array of options and with the deadline hours away, I stood on a Boston street corner in the bright light of a spring afternoon, trying to determine where my destiny lay. Earlier in the day, I had happened upon the Peabody Museum and wandered through its dusty halls, past dioramas of waxen figures dressed impeccably in the costumes of another time: Sioux warriors in full regalia, Haida women clothed in cedar bark robes, Huichol shaman enveloped in all the colors of the rainbow. With these images still swirling in my mind, I was approached on the street by another freshman, an intriguing character whom I had met just days before. When he mentioned his intention to study ethnology, my fate was somehow sealed. In the morning, I signed on as a student of social anthropology.

Sentiment alone, however, did not prepare me for what lay ahead. Within weeks, I fell into the orbit of Professor David Maybury-Lewis, who became my tutor. A man of searing intelligence, whose formal eloquence masked a deeply humane spirit, Maybury-Lewis remains one of the great Americanists, a brilliant scholar who had lived for years among the Akwé-Xavante and Xerente Indians in central Brazil. A student of Rodney Needham at Oxford, he had come to anthropology after earning a degree at Cambridge in Romance languages. His German, Russian, Danish, Spanish, and Portuguese were flawless, but it was the way he spoke English that fired the senses. His accent implied erudition. Combined with the precision of his thoughts, the effect was mesmerizing.

Maybury-Lewis had traveled to central Brazil in the mid-1950s to investigate and, in a sense, celebrate the so-called Gê anomaly. Throughout the nineteenth century and well into the first decades of the twentieth, anthropologists had maintained that technological sophistication and material well-being were a direct measure of the complexity of a culture, a convenient concept that invariably placed Victorian England at the top of a Darwinian ladder to success. Modern ethnographers rejected the notion out of hand, arguing that every human culture had, by biological definition, the same mental acuity. Whether this potential was realized through technological prowess or by the elaboration of intensely complex threads of memory inherent in a myth was a matter of cultural choice and historical circumstance. Nowhere was this modern notion more perfectly displayed than among the peoples of eastern Brazil, the complex of fierce tribes known as the Gê.

Living in the forests of Mato Grosso and on the arid savannas and uplands that separate the southern Amazon basin from the Atlantic coast, the Akwé-Xavante, Xerente, Kayapó, Timbira, and a host of other peoples all spoke dialects of the Gê language family. Semi-nomadic hunters and gatherers, they ranged across vast territories, hunting peccaries for meat, birds for brilliant plumage to be woven into ceremonial coronas that shone like the sun. Their material culture was exceedingly rudimentary. They knew nothing of canoes, though rivers dissected their lands, and as late as the 1950s were still dependent on the bow and arrow. They cleared the savanna, but their harvests were meager, their sparse plantings reminiscent of the dawn of agriculture.

Yet beneath the primitive veneer lay an astonishingly rich and complex worldview, a tangle of religious beliefs and myths that informed all of life and gave rise to patterns of social organization Byzantine in their sophistication yet perfectly elegant in their elaboration. This apparent contradiction of a marginal people, technologically backward yet mentally and intellectually afire, confounded many early ethnographers. Hence, the notion of the Gê anomaly. But the great French anthropologist Claude Lévi-Strauss saw the situation quite differently. For him, a scholar of immense vision capable of embracing all of the Americas in a single thought, the cultures of the Gê represented nothing more than a simple triumph of the human spirit and imagination.

Having lived briefly among the Bororo, a people whose social structure was very closely related to that of the Gê tribes, Lévi-Strauss had come to see their world as a universe of oppositions: man and woman, light and darkness, good and evil, the sun and the moon, the raw and the cooked, the wild and the tame. Every facet of their society—every ritual and institution, every concept of kinship and procreation, the very cycles of life and death, the transitions of birth and initiation, even the architecture and settlement pattern of the seasonal encampments—reflected a subconscious and indeed conscious attempt to resolve in harmony these opposing elements. Thus, there were two moieties, and two exogamous patrilineages entwined by cross-cousin marriages. The men lived apart from their families, in a ceremonial structure at the center of the enclosure. The women dwelled with the children in huts on the periphery of the encampment, at the edge of the human realm. Outside the domestic space was a ring of fields, feral and unkempt, where women birthed and husbands coupled with their wives. Beyond the clearings, the forest and savanna reached in all directions to the horizon.

This quest for balance, Lévi-Strauss maintained, was a fundamental human urge, a key adaptive trait that allowed peoples such as the Bororo to come to terms with the fragility of their lives and the harshness of the natural world that surrounded them. At the very least, it provided an illusion of control, that in their scattered encampments they were not utterly at the mercy of the fickle forces of life and death. Modern industrialized society has precisely the same need to insulate the individual from nature and indulges in similar illusions that it can be accomplished. The difference lies in the medium. We build machines and dwell in cities. The Gê peoples find protection in a web of ideas, beliefs, and ritual practices dreamed into being at the beginning of time.

From these insights, Lévi-Strauss elaborated a model of dualistic societies so simple and yet so all-encompassing as to almost defy belief. Clearly, it allowed for a better understanding of the Bororo. But what of the other societies of Central Brazil? It was, in part, a desire to challenge Lévi-Strauss, or at least to test his model, that in 1957 led David Maybury-Lewis to travel up the Río das Mortes, the River of the Dead, and to fly into a remote mission adjacent to the lands of the Akwé-Xavante, at the time the most feared and warlike of all the indigenous tribes of Brazil.

"From the air," Professor Maybury-Lewis recalled one afternoon when we met for a tutorial in his book-lined office overlooking the courtyard of the Tozzer Library, "everything looked right."

Laid out before him, just as Lévi-Strauss had described, were the men's circle and ritual shelters at the epicenter, the concentric rings of houses, fields, and forest. But after months on the ground, and despite having mastered the language, Maybury-Lewis was more confused than ever. There were not two patrilineages, but three, and the marriage rules were inconsistent with the simple bilateral pattern that Lévi-Strauss had reported for the Bororo. What's more, the bonds of kinship were crosscut by age sets.

Irrespective of lineage, every male of the same age shared the same age set. Each spanned five years, and there were eight in all. Thus, all boys aged five to ten, for example, or men thirty to thirty-five, were united as members of a named age set. Those few who lived beyond the age of forty once again became members of the first cohort. Several times a year, the age sets would divide into two teams for a race, with cohorts 1, 3, 5, and 7 going up against 2, 4, 6, and 8, an arrangement that ensured that each side would have a similar mix of infants, boys, men, and elders. The race itself entailed each side carrying a large and heavy log for miles across the savanna, a marathon of dust, sweat, and endurance that left every participant spent and exhausted.

Maybury-Lewis loved the excitement and avidly took part, though once again he found the ritual confusing. For one thing, no one was particularly concerned that the logs carried by each side be of similar weight. For another, it was not uncommon for the leading side to pause in the midst of the race, allowing the other to catch up. The first time Maybury-Lewis ran, his team did well, crossing the finishing line hours ahead of the opposition. He reveled in the victory, until he noticed that all of his teammates were downcast. The next time they raced, several weeks later, the other side won decisively, and everyone seemed crestfallen. Totally bewildered, Maybury-Lewis took part in yet a third race. This time, to the disappointment of the competitor in him, the sides approached and crossed the line at the same instant. To his utter surprise, both teams and the entire community erupted in a whirlwind of celebration.

"The goal wasn't to win," he recalled with a smile, "it was to arrive together."

All the tensions inherent in three competing lineages, every conflict within the culture, are distilled into two opposing factions, the two teams that give their all in a frenzied effort to reach a tie. Opposition and harmony, the resolution of conflict in ritual balance. It was more complex than Lévi-Strauss had ever imagined. The dualistic notion permeated every aspect of the culture, as did the central quest for resolution and equilibrium.

"What would happen," Maybury-Lewis asked me, "should the race not occur, or should it never end in a tie?"

I hesitated, uncertain how to answer. Then I heard myself say, "The culture would atrophy."

"Yes!" he exclaimed triumphantly.

To this day, I am not certain how I came up with that answer. Much of the content of what we had been discussing, patterns of kinship and marriage rules, subtleties of social organization, lay beyond me. But the log races I could understand, and intuitively I grasped their significance. For the first time, I understood the lesson of anthropology. I saw that as a people the Akwé-Xavante were profoundly different; but more importantly, I came to understand that those differences held the key to their cultural survival.

~

AS A YOUNG ANTHROPOLOGIST I never understood how I was supposed to turn up at some village—perhaps of the Barasana, a people of the Anaconda, who believed that their ancestors had come up the Milk River of the Amazon from the east only to be disgorged from the belly of the snake onto the banks of the upper affluents—and announce that I was staying for a year, and then notify the headman that he and his people were to feed and house me while I studied their lives. If someone that intrusive appeared on our doorsteps, we would call the police.

I learned, instead, to seek the proper conduit to a culture, the most appropriate means or metaphor to break down the inherent barrier that exists between a stranger and a people with whom that outsider finds himself living as a guest. In the Northwest Amazon, for example, and along the eastern flank of the Peruvian and Bolivian Andes, in the cloud forests of the Kamsa and Inga, and among lowland tropical peoples as diverse as the Chimane and Machiguenga, Kofán and Cubeo, the obvious vehicle was the botanical realm. These, after all, were societies that existed because of their plants. Their basic food was bitter manioc, a poisonous root rendered edible by women using a complex process mediated by ritual and infused with myth. From the astringent bark of lianas, their hunters extracted poisons that could kill, as well as potions and stimulants that conquered sleep, allowing men to move by night in the shadows of jaguar. Their shaman listened and heard voices, plant songs that provided clues to hidden pharmacological properties that, once exploited, allowed them to journey in trance to the stars. Plants fed the children, healed the elders, vanquished enemies. I became an ethnobotanist because I could not imagine any better way of understanding the lives of the people of the forest.

In the Canadian north, by contrast, in a world where animals dominate and the dialogue is between predator and prey, the central metaphor is the hunt. Unless one is able to follow caribou over the tundra, track moose through the forest, one can never fully embrace the rhythm of the culture. To record the myths of Athabaskan elders, one has to become a hunter, for the myths are an expression of the covenant that exists between men, women, and the wild, a way for the Indian people to rationalize the terrible fact that in order to live, they must kill the creatures they love most, the animals upon which they depend. Like so many lessons of anthropology, this was something that I learned through experience, living amongst a people, frequently making mistakes but always paying attention to the consequences.

In northwestern British Columbia, a year or so after graduating from college, I was hired as one of the first park rangers in the Spatsizi wilderness, a roadless track of some two million acres in the remote reaches of the Cassiar Mountains. The job description was deliciously vague: wilderness assessment and public relations. In two long seasons, our ranger team, myself and one other, Al Poulsen, a six-foot four vegetarian who grazed through meals and could conjure golden eagles out of the wild, encountered perhaps a dozen visitors. Wilderness assessment was a license to explore the park at will, tracking game and mapping the horse trails of outfitters, describing routes up mountains and down rivers, recording what we could of the movements of large populations of caribou and sheep, mountain goats, grizzly bears, and wolves.

In the course of these wanderings, we came upon an old Native gravesite on an open bench overlooking Laslui Lake, near the headwaters of the Stikine River. The wooden tombstone read, simply, "Love Old Man Antoine died 1926." Curious about the grave, I crossed the lake to the mouth of Hotleskwa Creek, where the Collingwood brothers, the outfitters for the Spatsizi, had established a spike camp. There, I found Alex Jack, an old Gitxsan man who had lived in the mountains most of his life. His Native name was Atehena, "he who walks leaving no tracks." Not only did Alex know of the grave, his own brother-in-law had laid the body to rest in it. Old Man Antoine, it turned out, was a legendary shaman, crippled from birth but possessed of the gift of clairvoyance. Alex had walked overland from his home at Bear Lake in the Skeena, a hundred and fifty miles to the south, in order to meet Antoine, only to arrive on the day of his death.

Intrigued by this link between a living elder, raised in seasonally nomadic encampments, totally dependent on the hunt, and a shaman born in the previous century who read the future in stones cast into water held in baskets woven from roots, I left my job as a park ranger and went to work with Alex. As we wrangled horses, repaired fences, guided the odd hunter in search of moose or goat, I would ask him to tell me the stories of the old days, the myths of his people and his land. He happily told tales of his youth, of the hunting forays that brought meat to the village and of the winter trading runs by dogsled to the coast, but he never said a word about the legends.

Long after I had given up on hearing the origin myths, I went out one morning to salvage a moose carcass abandoned by a trophy hunter. When I returned after a long day with a canoe full of meat, Alex was waiting for me. As we walked back across the meadow with our loads, he said very quietly that he remembered a story and invited me to drop by his tent later in the evening. To this day I do not know whether Alex had simply achieved a certain level of trust, or whether I had finally inquired about the stories in the correct manner, or whether the gift of meat had some greater significance. But that night, I began to record a long series of creator tales of We-gyet, the anthropomorphic figure of folly, the trickster/transformer of Gitxsan lore.

They were almost all whimsical stories of moral gratitude played out against and within the backdrop of nature. We-gyet, for example, eager to eat, swims beneath a gathering of swans and greedily grabs their legs, only to be dragged from the water as the flock rises in response, soaring toward the sun. Stranded in the sky, he lets go and comes crashing back to the earth, the force of the impact imbedding him in granite. A lynx comes by, and We-gyet, using his charm and guile, persuades the cat to lick away the rock. We-gyet rewards his savior with the tufts of hair that have since that time decorated the ears of every lynx.

To kill a grizzly, We-gyet takes advantage of the creature's pride. Moving with the speed of the wind, he flies past a berry patch, astonishing the bear with his grace and movements. The grizzly looks up, only to see We-gyet race by once more. After three passes, We-gyet stops, breathlessly approaches his prey, and collapses with laughter as he points disparagingly at the bear's testicles. "No wonder you can't run," he comments, "with those things dangling between your legs. I cut mine off years ago. See?" We-gyet has stained his groin with the blood-red sap of a willow. The grizzly, eager to remain the dominant creature in the forest, slices off his genitals and promptly bleeds to death.

Animals large and small featured in each of Alex's tales.

A hunting party away from home for many days grows tired, the young boys restless and bored. To pass the time, they cast a squirrel into their fire, a cruel gesture repeated again and again until the creature, unable to escape, disappears in the flames. The following morning, the hunters awake to find themselves camped in a circle at the base of an enormous cylinder of rock that reaches to the heavens, bluffs on all sides, no escape. Perplexed, a warrior tosses a pack dog into the fire, and to his surprise, the animal appears at the top of the rock face. One by one, each hunter slips into the flames and materializes alongside the dog, thus miraculously escaping the trap. They head for home, but when they enter their village and approach their loved ones, no one sees them. They reach out and try to touch their wives and mothers. Their hands pass through the bodies, like air. They are all dead, ghosts empty of will, punished for the crime of having, as Alex put it, "suffered that small squirrel."

Darkness is the time for stories, and in the glow of a kerosene lamp, with wind and rain falling upon the canvas, the tent that first night took on the warmth of a womb. Alex's words themselves had a certain magic, a power to influence not only the listener but the land itself, the very moment in time. When he told a story, he did not, as we might do, recount an anecdote, which by definition is a literary device, an abstraction, the condensation of a memory extracted from the stream of experience, a recollection of facts strung together with words. Alex actually lived the story again and again, returning in body and soul, in physical gesture and nuance, to the very place and time of its origin. At first, I thought this merely charming, and only after many years of listening, often to the same account told in the same way time and again, did I understand the significance of what he was sharing.

Alex did not come from a tradition of literacy. He had never learned to read or write with any degree of fluency. His soul had not been crushed by the priests in the residential schools. For most of his adult life, he had been a seasonally nomadic hunter, and his very vocabulary was inspired by the sounds of the wild. For him, the sweeping flight of a hawk was the cursive hand of nature, a script written on the wind. As surely as we can hear the voices of characters as we read the pages of a novel, so Alex could hear in his mind the voices of animals, creatures that he both revered and hunted. Their meat kept him alive. Their brains allowed skins to be worked into leather for moccasins and clothes, packsacks and the traces of his sled, the scabbard for his 30.06 rifle. Their blood could be cooked, the marrow of their bones sucked out and fed as a delicacy to children.

When Alex told a story, he did so in such a way that the listener actually witnessed and experienced the essence of the tale, entering the narrative and becoming transfixed by all the syllables of nature. Every telling was a moment of renewal, a chance to engage through repetition in the circular dance of the universe.

Alex never spoke ill of the wind or the cold. When hunting, he never referred to the prey by name until after the kill; then, he spoke directly to the animal with praise and respect, admiration for its strength and cleverness. His grandmother was Cree, people of the medicine power, who believe that language was given to humans by the animals. His mother was Carrier. In 1924, two years before Alex left Bear Lake to walk overland to the Stikine, an elder from the Bulkley Valley, quite possibly one of Alex's relatives, revealed something of the Carrier world to the anthropologist Diamond Jenness:

> We know what the animals do, what are the needs of the beaver, the bear, the salmon and other creatures, because long ago men married them and acquired this knowledge from their animal wives. Today the priests say we lie, but we know better. The white man has only been a short while in this country and knows very little about the animals; we have lived here thousands

> of years and were taught by the animals themselves. The white man writes everything down in a book so that it will not be forgotten; but our ancestors married animals, learned all their ways, and passed on this knowledge from one generation to another.

I did indeed write down Alex's tales, transcriptions of dozens of hours of conversations recorded intermittently over twenty-five years, committed to paper a few years before his death. Only after it was done did I realize that in a sense I had committed a form of violence, a transgression that bordered on betrayal. Extracted from the theater of his telling, the landscape of his memory, the sensate land, and the sibilant tones of the wild, the stories lost much of their meaning and power. Transposed into two dimensions by ink and paper, trapped on the page, they seemed childlike in their simplicity, even clumsy in their rhetoric.

But, of course, these stories were not meant to be recorded. They were born of the land and had their origins in another reality. Some time after I first learned of We-gyet from Alex, I asked him how long it took to tell the cycle of tales. He replied that he had asked his father that very question. To find out, they had strapped on their snowshoes in March, a time of good ice, and walked the length of Bear Lake, a distance of some twenty miles, telling the story as they went along. They reached the far end, turned, and walked all the way back home, and the story, Alex recalled, "wasn't halfway done."

In order to measure the duration of a story, the length of a myth, it was not enough to set a timepiece. One had to move through geography, telling the tale as one proceeded. For Alex and his father, this sense of place, this topography of the spirit, at one time informed every aspect of their existence. When, at the turn of the century, a Catholic missionary arrived at their village at Bear Lake, Alex's father was completely confounded by the Christian notion of heaven. He could not believe that anyone could be expected to give up smoking, gambling, swearing, carousing, and all the things that made life worth living in order to go to a place where they did not allow animals. "No caribou?" he would say in complete astonishment. He could not conceive of a world without wild things.

Alex lived for more than 90 years; his wife Madeleine reached 103, passing away a few seasons before Alex followed her to the grave. A year before he died, Alex gave me a small gift, a tool carved from caribou bone. Smooth as marble, though stained from years of use, it fit perfectly in my hand, the rounded and slightly serrated spoon-like tip protruding neatly from between finger and thumb. I had no idea what it might have been used for. Alex laughed. He had carved it more than eighty years before, following the lead of his father. It was a specialized instrument, used to skin out the eyelids of wolves. Only later did I realize that the eyelids in question were my own, and that Alex, having done so much to allow me to see, was, in his own way, saying good-bye.

~

Perhaps because I never knew my grandparents, who died before I was born, I have always been drawn to elders, enchanted by the radiance of men and women who have lived through times that I can only imagine: an old schoolmaster who scrambled out of the trenches on the first day of the Somme; a family doctor who treated the wounded along the partition line between India and Pakistan, when rivers of blood divided the Raj; Waorani shaman who knew the Amazonian forests before the arrival of missions. I am enticed by their memories, and, in a culture notably bereft of formal modes of initiation, I find comfort in their advice. From men like Alex, I have learned of a world without form, infused with spirit and prayer. But equally important to me is the landscape of the concrete, the formal realm of science.

Alex Jack at Ealue Lake, Stikine Valley,
British Columbia, 1996

A woodland caribou, Spatsizi wilderness,
British Columbia, 1979

Moose crossing a slough, Copper River Delta,
Alaska, 1995

A Swedish hunter at Hotleskwa Creek,
Spatsizi wilderness, British Columbia, 1980

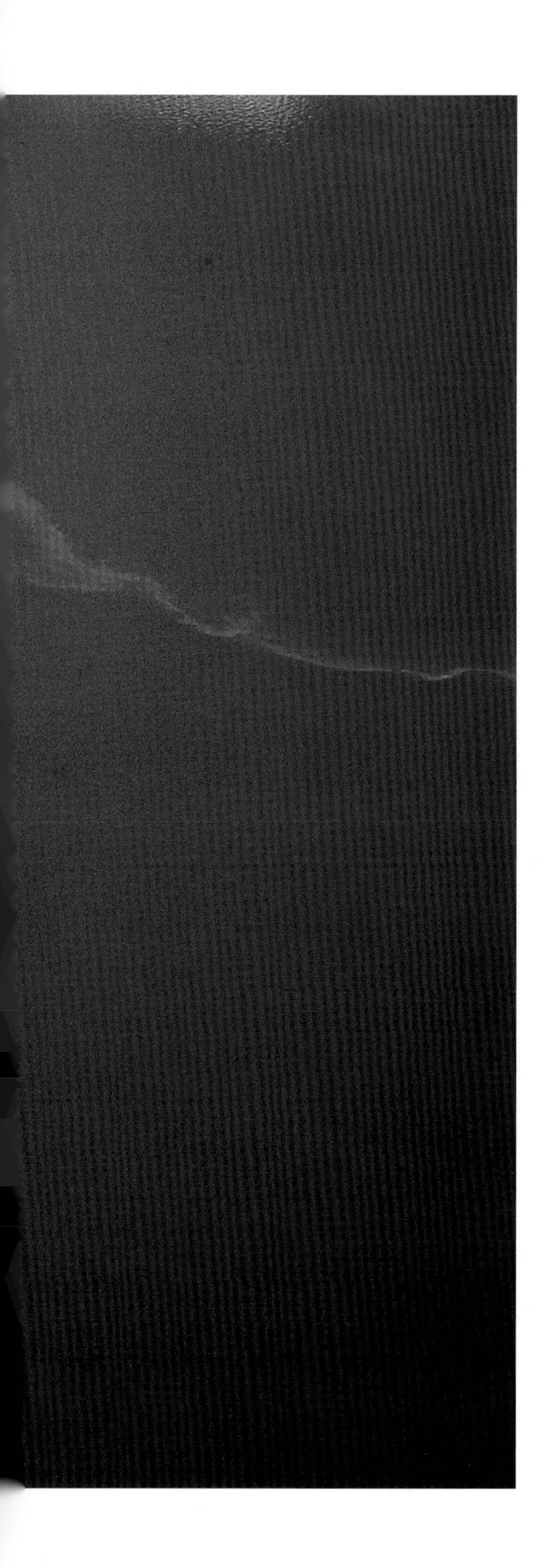

A red-tide bloom, Behm Canal,
southeast Alaska, 1995

In the early 1970s, a time of few heroes, there was one man who loomed large over the Harvard campus, Richard Evans Schultes, a kindly professor who demanded nothing but devotion to knowledge. In time, mountains in South America would bear his name, as would national parks. Prince Philip would call him "the father of ethnobotany." Students knew him as the world's leading authority on medicinal and hallucinogenic plants, the plant explorer who had sparked the psychedelic era with the discovery of psilocybin mushrooms in Mexico in 1938. Three years later, having proved that *teonanacatl*, the flesh of the gods, was indeed a mushroom, and having identified *ololiuqui*, the serpent vine, the second of the elusive Aztec hallucinogenic plants, Schultes turned his imagination to the forests of South America. Taking a semester's leave of absence from the university, he disappeared into the Northwest Amazon, where he remained for twelve years, mapping uncharted rivers and living among two dozen indigenous tribes, all the while in pursuit of the mysteries of the rain forest. He collected over twenty-seven thousand botanical specimens, including two thousand medicinal plants and over three hundred species previously unknown to science. For his students, he was a living link to the great naturalists of the nineteenth century and a distant era when the tropical rain forests stood immense, inviolate, a mantle of green stretching across entire continents.

By the time I met Schultes in the fall of 1973, it had been some years since he had been capable of active fieldwork. I found him at his desk in his fourth floor aerie in the Botanical Museum, dressed conservatively, peering across several large stacks of dried herbarium specimens. Introducing myself as one of his undergraduate students, I mentioned that I was from British Columbia and that I wanted to go to the Amazon and collect plants, just as he had done so many years before. The professor looked up from his desk and, as calmly as if I had asked for directions to the local library, said, "Well, when would you like to go?" A fortnight later I left for South America, where I remained during that first sojourn for fifteen months.

There was, of course, method in Schultes's casual manner. He took for granted the capacity of anyone to achieve anything. In this sense he was a true mentor, a catalyst of dreams. Though not by nature a modest man, he shared his knowledge and experience with his acolytes as naturally as a gardener brings water to a seed. Sometimes his faith in a student would lead to disappointment, but not often. His own achievements were legendary, and merely to move in his shadow was to aspire to greatness.

In Schultes, I found the perfect complement to Maybury-Lewis, my anthropology tutor. Whereas Maybury-Lewis awakened the soul through the sheer power of his intellect, the wonder of his words and ideas, Schultes inspired by the example of his deeds. In all the years I was formally his student, we never had an intellectual conversation. It was not his style. He was a true explorer, and the very force of his personality gave form and substance to the most esoteric of ethnobotanical pursuits. He would pass along these thoughts that were both gifts and challenges. "There is one river that I would very much like you to see," he would say, knowing full well that the process of getting to that river would involve experiences guaranteed to assure that were you able to reach the confluence alive, you would emerge from the forest a wiser and more knowledgeable human being.

Typical of the way Schultes operated was his suggestion, offered casually just before I left for South America, that I look up one of his former graduate students, Tim Plowman, in Colombia. Tim, who would become a close friend, was Schultes's protégé, and the professor had secured for him from the U.S. government the dream academic grant of the early 1970s, $250,000 to study a plant known to the Inca as the Divine Leaf of Immortality, the most sacred medicine of the Andes, coca, the notorious source of cocaine.

It was a remarkable assignment. Though the drug had long been the focus of public concern and hysteria, and efforts to eradicate the coca fields had been underway for nearly fifty years, astonishingly little was known about the actual plant. The botanical origins of the domesticated species, the chemistry of the leaf, the pharmacology of coca chewing, the plant's role in nutrition, the geographical range of the domesticated species, the relationship between the wild and cultivated species—all these were mysteries. No concerted effort had been made to document the role of coca in the religion and culture of the Andean and Amazonian Indians since W. Golden Mortimer's classic *History of Coca*, published in 1901. Tim's mandate from the government, made deliberately vague by Schultes, was to travel the length of the Andean Cordillera, traversing the mountains whenever possible, to reach the flanks of the *montaña* to locate the source of a plant that had inspired an empire. Eventually, Tim and I would spend over a year on the road, a journey made possible by the great professor and infused at all times with his spirit.

We knew, of course, that coca was the most revered plant of the Andes. The Inca, unable to cultivate the bush at the elevation of the imperial capital of Cusco, replicated it in fields of gold and silver that colored the landscape. No holy shrine in the land could be approached unless the supplicant had a quid of coca in his mouth. No field could be planted, no child brought into being, no elder released to the realm of the dead unless the transition was mediated with an offering of coca leaves for Pachamama, the Goddess of the Earth. To this day in parts of the Andes, distances are measured not in miles but in coca chews. When Runa people meet, they do not shake hands, they exchange leaves. Soothsayers divine the future by interpreting the patterns of leaves cast onto cloth and the patterns in the venation of the leaves, a skill that can only be possessed by someone who has survived a lightning strike.

In time, Tim would solve the botanical mystery, identify the point of origin of the domesticated species, and reveal how they had diverged through centuries as their cultivation had spread over much of a continent. But perhaps his greatest research contribution came about from a simple nutritional analysis, the results of which horrified his government backers, even as they transformed scientific thinking about this most sacred of plants. Coca leaves do contain a small amount of cocaine, but only about as much as there is caffeine in a coffee bean. When the leaves are chewed, the drug is absorbed slowly through the mucous membrane of the mouth; it is a benign and useful stimulant in a harsh and unforgiving landscape. Highly effective as a treatment for altitude sickness, the leaves proved also to be extraordinarily nutritious. Rich in vitamins, coca has more calcium than any plant ever assayed by the U.S. Department of Agriculture, suggesting a vital role in a diet that traditionally lacked dairy products, especially for nursing mothers. It was also suggested that the plant enhances the ability of the body to digest carbohydrates at high elevation, again an ideal complement for a diet based on potatoes. In one elegant scientific assay, Tim revealed that coca was not a drug but a sacred food, a medicinal plant that had been used without any evidence of toxicity, let alone addiction, for over four thousand years by the peoples of the Andes. This revelation put into stark profile the draconian efforts underway then and continuing to this day to eradicate the traditional fields with herbicides that poison the myriad streams cascading out of the mountains to form the headwaters of the Rivers Amazon.

~

Coca was the lens through which the ancient rhythms and patterns of life in the Andes gradually came into focus. Wherever Tim and I traveled, we encountered evidence of worlds that had never been vanquished, indigenous communities that despite desperate struggles remained inextricably linked to their homelands. Nowhere was the spirit of survival stronger than among the Ika and Kogi, descendants of an ancient civilization that had flourished on the Caribbean plain of Colombia for five hundred years before the arrival of Europeans. Since the time of Columbus, these Indians have resisted invaders by retreating higher and higher into the inaccessible reaches of the Sierra Nevada de Santa Marta, the highest coastal mountain range on Earth. Ruled to this day by a ritual priesthood, they consider themselves the Elder Brothers. We, who to their minds have ruined much of the world, are deemed the Younger Brothers.

Tim and I entered the mountains from the south, along a narrow track that rose in a day and a night through cactus and thorn scrub to a steep river draw carved into the rising flank of the massif. Just after dawn, having made our way through the shadowy darkness, past scattered houses of stone linked one to another by small fields of coca, leaves translucent in the early morning light, we came upon a portal to the sun. Framed within its arch was a solitary figure, a silhouette blocking entry to the upper valley of a river known to the Indians as the Donachuí.

His name was Adalberto Villafañe. He was a young man, perhaps twenty, strikingly handsome with fine features and black hair flowing down past his shoulders. He wore a white cotton cloak held at the waist by a belt of fiber. His leggings were of the same rough cotton. His sandals had been cut from a rubber tire and, together with his fez-like hat of woven sisal, revealed that he was Ika. The Kogi, a more reclusive people, disdain the use of hats and shoes, and live higher in the mountains, closer to what they believe to be the heart of the world. Across each of Adalberto's shoulders hung a woven bag decorated with brilliant geometric designs. These contained coca. In his left hand was a small bottle-shaped gourd. A thick quid of the leaves created a bulge in his cheek. As we explained the purpose of our visit, he removed from the gourd a lime-coated stick, which he placed in his mouth. He bit down gently. A trickle of saliva ran past his lips as he withdrew the stick, now wet with coca. Reflexively, he began to rub the head of his gourd with the stick, a habit of years that had resulted in a crown of calcium carbonate, the lime of burnt seashells, built up around the top of the gourd, shaped carefully, a symbol of prestige, the measure of the man.

At the time, I did not know that Adalberto's gourd had been a gift from the priest who had officiated at his marriage. The lime is essential, for it makes potent the plant, adding alkali to saliva and thus facilitating the absorption of the small amount of cocaine within the leaves. When a man marries, the priest presents him with a perforated gourd and, before his eyes, makes love to the bride, thus suggesting through ritual the fundamental notion that as a man weds himself to a life of matrimony, fidelity, and procreation, so he weds himself to the destiny of the ancestors and a lifetime dedicated to the sacred leaves. Nor did I know that the Ika and Kogi societal ideal is to abstain from sex, eating, and sleeping while staying up all night, chewing leaves, and chanting the names of the ancestors. What I saw in the moment was a simple man, decent beyond words, who found satisfaction in our explanation and was willing to accompany us into his homeland.

Turning abruptly, Adalberto led us through the stone wall and along a chalky trail that ran through plantings of maize and coca to a settlement that had existed for untold generations, a cluster of stone huts in the shade of frail trees and a temple where the elders awaited. Each of them spoke in turn, and after some deliberation, their voices met in a decision to allow us to proceed.

~

As we moved about the mountains over the next fortnight, collecting plants by day, reading and talking with Adalberto and the elders by night, the patterns of life in the Sierra slowly came into focus. Much of what we learned came from the writings of the great Colombian anthropologist Gerardo Reichel-Dolmatoff, who had lived among the Kogi and Ika in the 1940s. Without his insights, gleaned over the course of months of fieldwork, the baroque world of the Sierra would have remained to us utterly incomprehensible. For beneath the veneer of everyday life, of fields being planted, crops being sold, children being taught, lay a complex of sacred laws and expectations, a body of beliefs astonishing in their complexity, profound in their implications, luminous in their potential.

According to Reichel-Dolmatoff, the Kogi and Ika draw their strength from the Great Mother, a goddess of fertility who dwells at the heart of the world in the snow and ice of the high Sierra, the destination of the dead and the source of the rivers and streams that bring life to the fields of the living. At the first dawning when the Earth was still soft, the Great Mother stabilized it by thrusting her enormous spindle into the center, penetrating the nine levels of existence. The Lords of the Universe, born of the Great Mother, then pushed back the sea and lifted up the Sierra Nevada around the world axis, thrusting their hair into the soil to give it strength. From her spindle, the Great Mother uncoiled a length of cotton thread with which she traced a circle around the mountains, circumscribing the Sierra Nevada, which she declared to be the land of her children. Thus, the spindle became a model of the cosmos. The disk is the Earth, the whorl of yarn is the territory of the people, the individual strands of spun cotton are the thoughts of the sun.

For the Indians of the Sierra, everything begins and ends with the loom, and the metaphor of thread in the cosmic cloth. Constantly on the move as they gather food and various resources, the Indians refer to their wanderings as "weavings," each journey a thread woven into a sacred cloak over the Great Mother, each seasonal movement a prayer for the well-being of the people and the entire Earth. When the people of the Sierra plant a field, the women sow lines of crops parallel to the sides of the plot. The men work their way across the field in a horizontal direction. The result, should the domains of man and woman be superimposed one upon the other, is a fabric. The garden is a piece of cloth.

The surface of the Earth itself is an immense loom upon which the sun weaves the fabric of existence. The Indians acknowledge this in the architecture of their temples, simple structures with high conical roofs supported by four corner posts. On the dirt floor, positioned between the central axis of the temple and each of the four posts, are four ceremonial hearths that represent the four lineages founded at the beginning of time. In the middle is a fifth hearth, representative of the sun.

The orientation of the temples and the hearths within them is precise and critical. On the summer solstice, as the sun rises above the mountains, a narrow beam of sunlight shines through a hole in the roof and falls on the hearth that lies in the southwest corner, then moves across the floor until, just before dusk, it reaches the hearth in the southeast corner. On the winter solstice, the beam of light passes through the hole in the roof to touch the northwest hearth in the morning and passes over the floor to strike the northeast hearth at dusk. On both the fall and spring equinoxes, the beam of light slices a path equidistant between north and south; with the sun high in the sky, the central hearth, the most sacred of the five, is bathed in a vertical column of light. At that moment, a waiting priest lifts a mirror to the sun; as the light of the Father fertilizes the womb of the living, the mirrored light forms a cosmic axis along which the prayers of the people may ascend to the heavens.

Thus, over the course of a year, the sun literally weaves the lives of the living on the loom of the temple floor. The strands of the warp are laid down on the solstice, and the cloth is completed on the equinox, at which time the priest begins to dance at the eastern door of the temple, slowly moving across to the western entrance, while in gesture and song drawing a rod behind him. On reaching the western door, the priest pulls forth the imaginary rod, and the fabric of the sun unfolds: a new cloth is dreamt into being, the divine weaver soars over the loom, and life continues.

For the Kogi, the equilibrium of the world spun into being at the beginning of time is completely dependent on the moral and spiritual integrity of the Elder Brothers. The goal of life is knowledge, not wealth. Only through insight and attention can one achieve an understanding of good and evil, and an appreciation of the sacred obligations that human beings have to the Earth and the Great Mother. With knowledge come wisdom and tolerance, though wisdom is an elusive goal. In a world animated by the sun's energy, people invariably turn for guidance to the sun priests, the enlightened *mámas* who control the cosmic forces through their prayers and rituals, songs and incantations. Though they rule the living, the mámas have no special privileges, no outward signs of prestige, but their pursuit of wisdom entails an enormous burden: the survival of the people and the entire Earth depends on their labors.

Those who are chosen for the priesthood through divination are taken from their families as infants and carried high into the mountains to be raised by a máma and his wife. For eighteen years, they are never allowed to meet a woman of reproductive age or to experience daylight, forbidden even to know the light of a full moon. They sleep by day, waking after sunset, and are fed a simple diet of boiled fish and snails, mushrooms, grasshoppers, manioc, squash, and white beans. They must never eat salt or food not known to the ancients, and not until they reach puberty are they permitted to eat meat.

The apprenticeship falls into two phases, each of nine years duration, symbolic of the nine months spent in a mother's womb. During the first phase, the apprentices learn songs and dances, mythological tales, the secrets of Creation, and the ritual language of the ancients. The second nine years are devoted to the art of divination, techniques of breathing and meditation that lift them into trance, prayers that give voice to the inner spirit. The apprentices pay little heed to the mundane tasks of the world, but they do learn everything about the Great Mother, the secrets of the sky and the Earth, the wonder of life itself in all its manifestations. Knowing only darkness and shadows, they acquire the gift of visions and become clairvoyant, capable of seeing not only into the future and past but through all the material illusions of the universe. In trance, they can travel through the lands of the dead and into the hearts of the living.

Finally, after years of study and rigorous practice, of learning of the beauty of the Great Mother, of honoring the delicate balance of life, of appreciating ecological and cosmic harmony, a great moment of revelation arrives. On a clear morning, with the sun rising over the flank of the mountains, the apprentices are led into the light of dawn. Until then, the world has existed only as a thought. Now, for the first time, they see the world as it is, in all its transcendent beauty. Everything they have learned is affirmed. Standing at their side, the máma sweeps an arm across the horizon as if to say, "You see, it is as I told you."

WHEN I FIRST TRAVELED down the spine of the Andean Cordillera, past the remnants of temples and enormous storehouses that once fed armies in their thousands, through valleys transformed by agricultural terraces, past narrow tracks of flat stones, all that remains of the fourteen thousand miles of roads that once bound the Inca Empire, it was difficult to imagine how so much could have been accomplished in less than a century. The empire, which stretched over three thousand miles, was the largest ever forged on the American continent. Within its

Ika girls at Sagrome, Sierra Nevada de Santa Marta, Colombia, 1977

Men returning home from working in the fields,
Chinchero, Peru, 1982

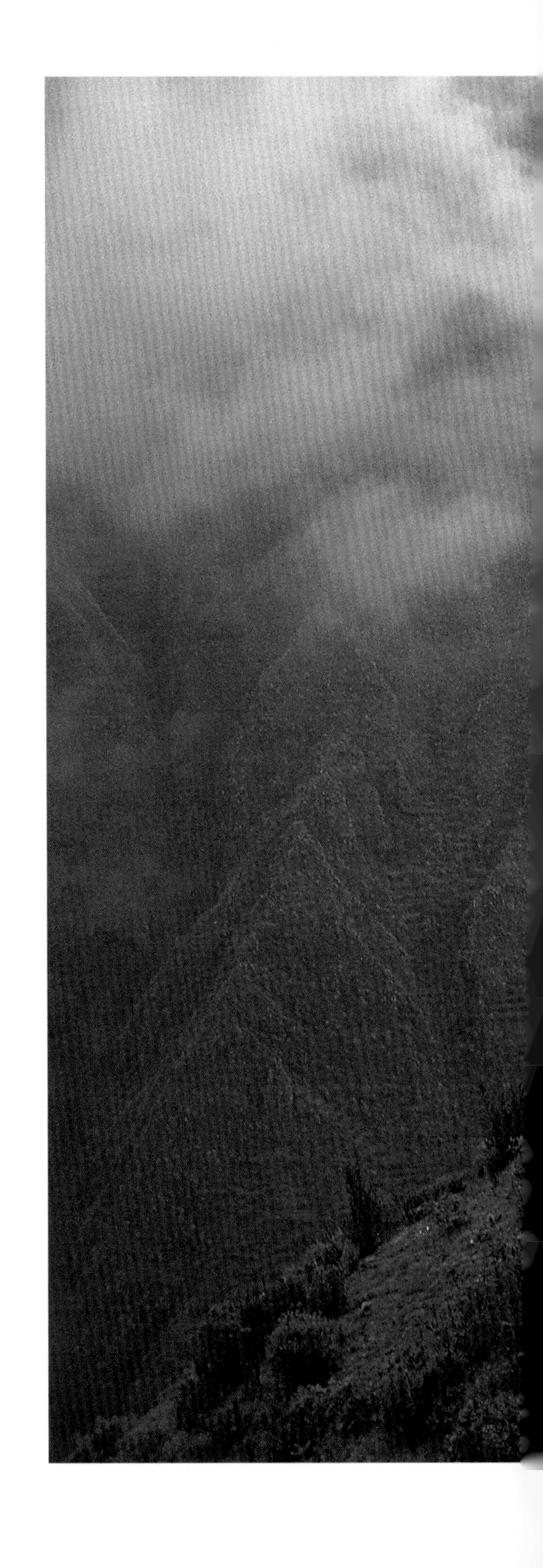

The Río Apurimac, headwaters of the Amazon,
Cordillera Vilcabamba, Peru, 1981

An elder, home from the fields, Chinchero, Peru, 1982

boundaries lived nearly all the people of the Andean world. There was said to be no hunger. All matter was perceived as divine, the Earth itself the womb of creation.

When the Spaniards saw the monuments of the Inca, they could not believe them to be the work of men. The Catholic Church declared the stonework to be the product of demons, an assertion no more fantastic than many more recent attempts to explain the enigma of Inca masonry as being of extraterrestrial origin. There was, of course, no magic technique. Only time, immense levies of workers, and an attitude toward stone that most Westerners find impossible to comprehend.

In the early spring of 1975, I visited the sacred valley of the Urubamba for the first time. As I walked the ruins of Pisac, a redoubt perched on a high mountain spur, I saw that the entire face of the massif was transformed by terraces which, when viewed from afar, took on the form of a gigantic condor, wings spread wide as if to protect the fortress. The river far below had been channeled by the Inca, lined for much of its length with stone walls to protect fields from flooding and to maximize agricultural production in one of the most fertile valleys of the empire. The faint remains of other terraces marked every mountainside. The entire landscape had been transformed, a stunning engineering feat for a people who knew nothing of the wheel and had no iron tools. As I sat in the Temple of the Moon, surrounded by some of the finest Inca masonry in existence, I recorded these notes in my journal and later incorporated them into the book *One River*:

> For the people of the Andes, matter is fluid. Bones are not death but life crystallized, and thus potent sources of energy, like a stone charged by lightning or a plant brought into being by the sun. Water is vapor, a miasma of disease and mystery, but in its purest state it is ice, the shape of snowfields on the flanks of mountains, the glaciers that are the highest and most sacred destination of the pilgrims. When an Inca mason placed his hands on rock, he did not feel cold granite; he sensed life, the power and resonance of the Earth within the stone. Transforming it into a perfect ashlar or a block of polygonal masonry was service to the Inca, and thus a gesture to the gods, and for such a task, time had no meaning. This attitude, once harnessed by an imperial system capable of recruiting workers by the thousand, made almost anything possible.
>
> If stones are dynamic, it is only because they are part of the land, of Pachamama. For the people of the Andes, the Earth is alive, and every wrinkle on the landscape, every hill and outcrop, every mountain and stream has a name and is imbued with ritual significance. The high peaks are addressed as Apu, meaning "lord." Together, the mountains are known as the Tayakuna, the fathers, and some are so powerful that it can be dangerous even to look at them. Other sacred places, a cave or mountain pass, a waterfall where the rushing water speaks as an oracle, are honored as the Tirakuna. These are not spirits dwelling within landmarks. Rather, the reverence is for the actual place itself.
>
> A mountain is an ancestor, a protective being, and all those living within the shadow of a high peak share in its benevolence or wrath. The rivers are the open veins of the Earth, the Milky Way their heavenly counterpart. Rainbows are double-headed serpents which emerge from hallowed springs, arch across the sky, and bury themselves again in the earth. Shooting stars are bolts of silver. Behind them lie all the heavens, including the dark patches of cosmic dust, the negative constellations which to the highland Indians are as meaningful as the clusters of stars that form animals in the sky.

These notions of the sanctity of land were ancient in the Andes. The Spanish did everything in their power to crush the spirit of the people, destroying the temples, tearing asunder the sanctuaries, violating the offerings to the sun. But it was not a shrine that the Indians worshipped, it was the land itself: the rivers and waterfalls, the rocky outcrops and mountain peaks, the rainbows and stars. Every time a Catholic priest planted a cross on top of an ancient site, he merely confirmed in the eyes of the people the inherent sacredness of the place. In the wake of the Spanish Conquest, when the last of the temples lay in ruins, the Earth endured, the one religious icon that even the Spaniards could not destroy. Through the centuries, the character of the relationship between the people and their land has changed, but not its fundamental importance.

ONE AFTERNOON NOT LONG AGO, in the small Andean town of Chinchero just outside of Cusco, I sat on a rock throne carved from granite. At my back was the sacred mountain Antakillqa, lost in dark clouds yet illuminated in a mysterious way by a rainbow that arched across its flank. Below me, the terraces of Chinchero fell away to an emerald plain, the floor of an ancient seabed, beyond which rose the ridges of the distant Vilcabamba, the last redoubt of the Inca, a landscape of holy shrines and lost dreams where Tupac Amarú waged war and the spirit of the Sun still ruled for fifty years after the Conquest. Two young boys played soccer on the village green, a plaza where once Topa Inca Yupanqui, second of the great Inca rulers, reviewed his troops. On the very stone where I rested, he, no doubt, had stood, for this village of adobe and whitewashed homes, this warren of cobblestones, mud and grass, had been built upon the ruins of his summer palace.

For four hundred years, the Catholic church, perched at the height of the ruins overlooking the market square, had dominated the site. A beautiful sanctuary, it bears today none of the scars of the Conquest. It is a place of worship that belongs to the people, and there are no echoes of tyranny. Within its soaring vault, in a space illuminated by candles and the light of pale Andean skies, I once stood at the altar, a newborn child in my arms, a boy swaddled in white linen, as an itinerant priest dripped holy water onto his forehead and spoke words of blessing that brought the infant into the realm of the saved. After the baptism, there was a celebration, and the child's parents, my new *compadres*, toasted every hopeful possibility. I, too, made promises, which in the ensuing years I attempted to fulfill. I had no illusions about the economic foundation of the bond. From me, my compadres hoped to secure support: in time, money for my godchild's education, perhaps the odd gift, a cow for the family, a measure of security in an uncertain nation. From them, I wanted nothing but the chance to know their world, an asset far more valuable than anything I could offer.

This pact, never spoken about and never forgotten, was, in its own way, a perfect reflection of the Andes, where the foundation of all life, both today and in the time of the Inca, has always been reciprocity. One sees it in the fields, when men come together and work in teams, moving between rows of fava beans and potatoes, season to season, a day for a day, planting, hoeing, weeding, mounding, harvesting. There is a spiritual exchange in the morning when the first of a family to awake salutes the sun, and again at night when a father whispers prayers of thanksgiving and lights a candle before greeting his family. Every offering is a gift: blossoms scattered onto fertile fields, the blessing of the children and tools at the end of each day, coca leaves presented to Pachamama at any given moment. When people meet on a trail, they pause and exchange *k'intus* of coca, three perfect leaves aligned to form a cross. Turning to face the nearest *apu*, or mountain spirit, they bring the leaves to their mouths and blow softly, a ritual invocation that sends the essence of the plant back to the earth, the community, the sacred places, and the souls of the ancestors. The exchange of leaves is a social gesture, a way of acknowledging a human connection. But the blowing of the *phukuy*, as it is called, is an act of spiritual reciprocity, for in giving selflessly to the earth, the individual ensures that in time the energy of the coca will return full circle, as surely as rain falling on a field will inevitably be reborn as a cloud.

ALMOST TWENTY YEARS after first visiting Chinchero, I returned to participate in an astonishing ritual, the *mujonomiento*, the annual running of the boundaries. Since the time of the Inca, the town has been divided into three *ayullus*, or communities, the most traditional of which is Cuper, the home of my compadres and, to my mind, the most beautiful, for its lands encompass Antakillqa and all the soaring ridges that separate Chinchero from the sacred valley of the Urubamba. Within Cuper are four hamlets, and once each year, at the height of the rainy season, the entire male population, save those elders physically incapable of the feat, runs the boundaries of their respective communities. It is a race but also a pilgrimage, for the frontiers are marked by mounds of earth, holy sites where prayers are uttered and ritual gestures lay claim to the land. The distance traveled by the members of each hamlet varies. The track I was to follow, that of Pucamarca, covers some fifteen miles, but the route crosses two Andean ridges, dropping a thousand feet from the plaza of Chinchero to the base of Antakillqa, then ascending three thousand feet to a summit spur before descending to the valley on the far side, only to climb once more to reach the grasslands of the high *puna* and the long trail home.

At the head of each contingent would dart the *waylaka*, the strongest and fleetest of the youths, transformed for the day from male to female. Dressed in heavy woolen skirts and a cloak of indigo, wearing a woman's hat and delicate lace, the waylaka would fly up the ridges, white banner in hand. At every boundary marker, the transvestite must dance, a rhythmic turn that like a vortex draws to the peaks the energy of the women left behind in the villages far below. Each of the four hamlets of Cuper has its own trajectory, just as each of the three ayullus has its own land to traverse. By the end of the day, all of Chinchero would be reclaimed: the rich plains and verdant fields of Ayullupunqu; the lakes, waterfalls, mountains, and cliffs of Cuper; the gorges of Yanacona, where wild things thrive and rushing streams carry away the rains to the Urubamba. Adversaries would have been fought, spirits invoked, a landscape defined, and the future secured.

This much I knew as I approached the plaza on the morning of the event. Before dawn, the blowing of the conch shells had awoken the town, and the waylakas, once dressed, had walked from house to house, saluting the various authorities: the *curaca* and *alcalde*; the officers of the church; and the *embarados*, those charged with the preservation of tradition. At each threshold, coca had been exchanged, fermented maize *chicha* imbibed, and a cross of flowers hung in reverence above the doorway. For two hours, the procession had moved from door to door, musicians in tow, until it encompassed all of the community and drew everyone in celebration to the plaza where women waited, food in hand: baskets of potatoes and spicy *piquante*, flasks of chicha, and steaming plates of vegetables. There I lingered, with gifts of coca for all. At my side was my godson, Armando. A grown man now, father of an infant girl, he had been a tailor but worked now in the markets of Cusco, delivering sacks of potatoes on a tricycle rented from a cousin. He had returned to Chinchero to be with me for the day.

What I could never have anticipated was the excitement and the rush of adrenaline, the sensation of imminent flight as the entire assembly of men, prompted by

CHICAGO HEIGHTS PUBLIC LIBRARY

some unspoken signal, began to surge toward the end of the plaza. With a shout, the waylaka sprang down through the ruins, carrying with him more than a hundred runners and dozens of young boys who scattered across the slopes that funneled downward toward a narrow dirt track. The trail fell away through a copse of eucalyptus and passed along the banks of a creek that dropped to the valley floor. A mile or two on, the waylaka paused for an instant, took measure of the men, caught his breath, and was off, dashing through thickets of buddleja and polylepis as the rest of us scrambled to keep sight of his white banner. Crossing the creek draw, we moved up the face of Antakillqa. Here, at last, the pace slowed to something less than a full run. Still, the men leaned into the slope with an intensity and determination unlike anything I had ever known. Less than two hours after leaving the village, we reached the summit ridge, a climb of several thousand feet.

There we paused, as the waylaka planted his banner atop a *mujon*, a tall mound of dirt, the first of the border markers. The authorities added their ceremonial staffs, and as the men piled on dirt to augment the size of the mujon, Don Jeronimo, the curaca, sang rich invocations that broke into a cheer for the well-being of the entire community. By this point, the runners were as restless as race horses, frantic to move. A salutation, a prayer, a generous farewell to those of Cuper Pueblo, another of the hamlets, who would track north, and we of Pucamarca were off, heading east across the backside of the mountain to a second mujon located on a dramatic promontory overlooking all of the Urubamba. Beyond the hamlets and farms of the sacred valley, clouds swirled across the flanks of even higher mountains, as great shafts of sunlight fell upon the river and the fields far below.

We pounded on across the backside of the mountain and then straight down at a full run through dense tufts of ichu grass and meadows of lupine and rue. Another mujon, more prayers, handfuls of coca all around, blessings and shouts, and a mad dash off the mountain to the valley floor, where, mercifully, we older men rested for a few minutes in the courtyard of a farmstead owned by a beautiful elderly woman who greeted us with a great ceramic urn of frothy chicha. One of the authorities withdrew from his pocket a sheet of paper listing the names of the men and began to take attendance. Participation in the mujonomiento is obligatory, and those who fail to appear must pay a fine to the community. As the names were called, I glanced up and was stunned to see the waylaka, silhouetted on the skyline hundreds of feet above us, banner in hand, moving on.

So the day went. The rains began in early afternoon, and the winds blew fiercely by four. By then nothing mattered but the energy of the group, the trail at our feet, and the distant slope of yet another ridge to climb. Warmed by alcohol and coca leaves, the runners fell into reverie, a curious state of joy and release, almost like a trance.

Darkness was upon us as we rushed down the final canyon on a broad muddy track where the water ran together like mercury and disappeared beneath the stones. Approaching the valley floor and the hamlet of Cuper Alto, where women and children waited, the rain-soaked runners closed ranks behind the waylaka to emerge from the mountains as a single force, an entire community that had affirmed through ritual its sense of place and belonging. In making the sacrifice, the men had reclaimed a birthright and rendered sacred a homeland. Once reunited with their families, they drank and sang, toasting their good fortune as the women served great steaming bowls of soup from iron cauldrons. And, of course, late into night, the waylakas danced.

EVERY CULTURE FACES the same fundamental challenges. Men and women come together, children are brought into

CHICAGO HEIGHTS PUBLIC LIBRARY

the world, nurtured, and sheltered; elders are led into the realm of death as fearlessly as the imagination allows. To be human is to know the terror and splendor of a night sky, the crush of storms, the blood cries of enemies sweeping in with the dawn. Such is our common experience. To bring order to chaos, sense to sensation, we have created rules, which cross-culturally are remarkable in their consistency. The behavior prescribed by the Ten Commandments, for example, would be readily endorsed by peoples throughout the world, because the rules work and allow us to survive as a social species. It is a rare society indeed that tolerates murder, thievery, the violation of established patterns of marriage and procreation. Without such laws, the thin veneer of civilization would be shattered, the veil of culture reduced to tatters.

Yet within this common fabric, this cloak of humanity, lie the individual threads of specific and highly specialized ways of life, distinct cultures, each with its unique and wondrous dream of the Earth. Unraveling the cloth and holding the strands to the light is the practice and contribution of ethnography.

The Kogi dream of the Earth, for example, as revealed by Gerardo Reichel-Dolmatoff, represents at least in part a societal and spiritual ideal, a metaphor linking the living and their everyday reality with the promise and hope of higher possibilities. The sacred laws of the Great Mother provide the Kogi with an image of a perfect way of being. A close examination of village life would no doubt turn up contradictions, for such is the nature of human behavior, but the Kogi remain remarkably true to the dictates of their religion. Ultimately what matters, however, is not how closely they follow a set of rules but rather what those rules say about how the Kogi perceive their place in the world. The full measure of a culture embraces both the actions of its people and the quality of their aspirations, the character and nature of the metaphors that propel them onward.

The significance of an esoteric belief lies not in its veracity in some absolute sense but in what it can tell us about a culture. Is a mountain a sacred place? Does a river follow the ancestral path of an anaconda? Do the prayers of the Kogi actually maintain the cosmic balance? Who is to say? What matters is the potency of the belief and the manner in which the conviction plays out in the day to day life of a people. A child raised to believe that a mountain is the abode of a protective spirit will be a profoundly different human being from a youth brought up to believe that a mountain is an inert mass of rock ready to be mined. A Kwakwa̲ka̲'wakw boy raised to revere the coastal forests of the Pacific Northwest as the abode of Huxwhukw and the Crooked Beak of Heaven, cannibal spirits living at the north end of the world, will be a different person from a Canadian child taught to believe that such forests exist to be logged.

Herein, perhaps, lies the essence of the relationship between indigenous peoples and the natural world. Life in the malarial swamps of New Guinea, the chill winds of Tibet, the white heat of the Sahara, leaves little room for sentiment. Nostalgia is not a trait commonly associated with the Inuit. Nomadic hunters and gatherers in Borneo have no conscious sense of stewardship for mountain forests that they lack the technical capacity to destroy. What these cultures have done, however, is to forge through time and ritual a traditional mystique of the Earth that is based not only on deep attachment to the land but also on far more subtle intuition—the idea that the land itself is breathed into being by human consciousness. Mountains, rivers and forests are not perceived as inanimate, as mere props on a stage upon which the human drama unfolds. For these societies, the land is alive, a dynamic force to be embraced and transformed by the human imagination. This sense of belonging and connection, noted by ethnographers working among traditional societies throughout the Andes, is also the invisible constant of the Amazon.

A traditional healer at prayer,
Chinchero, Peru, 1998

Children of Huilloc, Peru, 1998

During the ceremonial running of the boundaries,
the men of Pucamarca arrive at a sacred site
overlooking the Urubamba Valley,
Chinchero, Peru, 2001

An Ayni labor party at the first hoeing of the potato fields, Chinchero, Peru, 1982

Young girls spinning wool, Chinchero, Peru, 2001

A boy carrying a bundle of mustard greens to feed the animals, Chinchero, Peru, 2001

The Salineras of Maras, a source of salt in the Andes for a thousand years, Peru, 2001

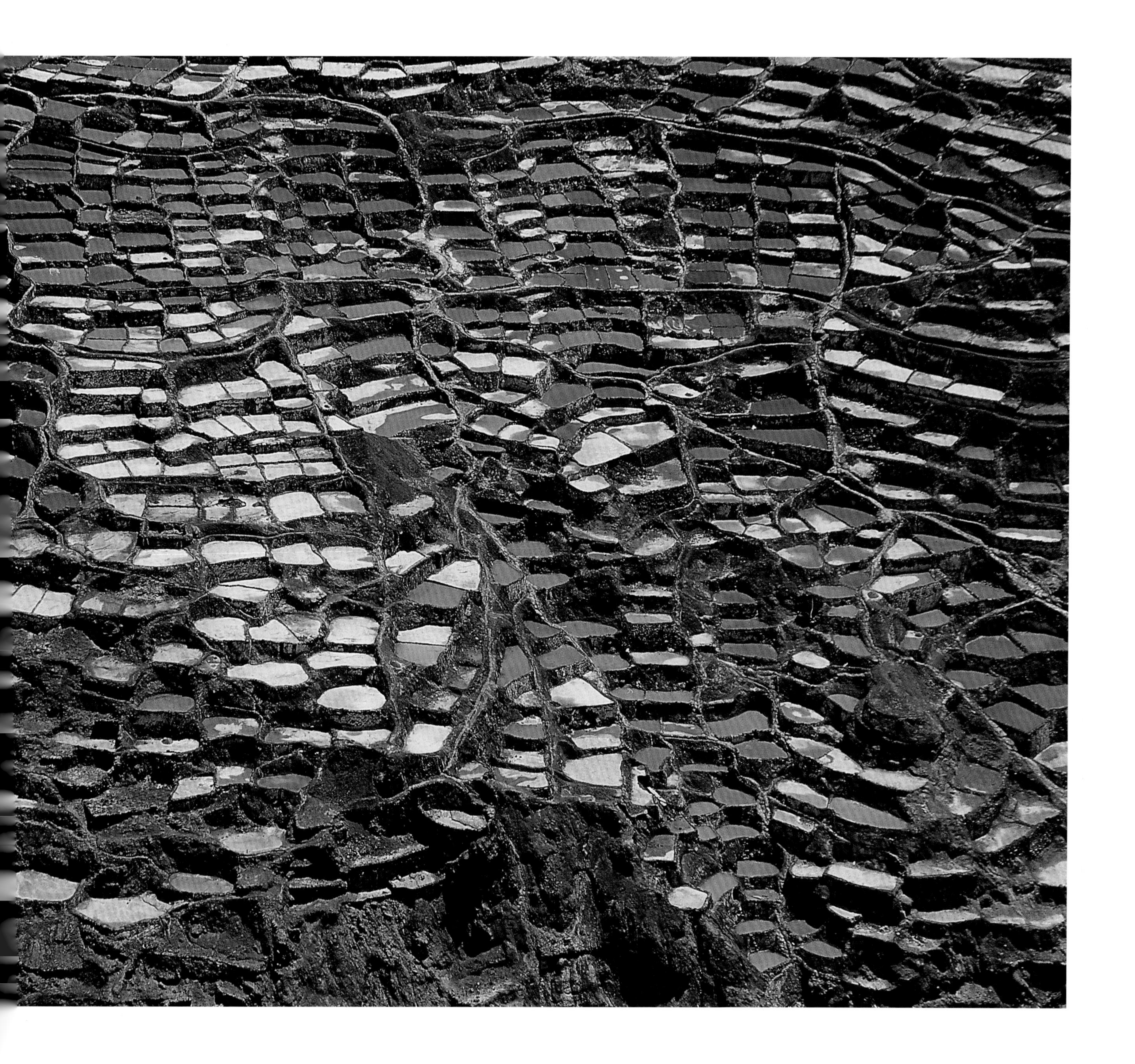

The Warao settlement of Winikina,
Orinoco Delta, Venezuela, 1997

CHAPTER THREE

THE FOREST AND THE STARS

If the Andes draw one to the light and shelter of a radiant sky, the tropical lowlands seduce with the promise of raw fecundity. Lie flat on the forest floor and see how long it takes to be colonized by fungi, tormented by insects. Put your mouth to the ground and breathe on a raiding column of army ants and see what happens. They are chemical beings, with eyes incapable of discerning images. They respond to odor. The scent of your breath excites their taste for flesh.

At night, the canopy comes alive with a familiar cacophony: frogs and cicadas, the roar of howlers, the unexpected bark of a jaguar. But at midday, the voices come from below, from the surface of the ground. Sit still and you can actually hear the crunching sound of long-horned beetles chewing through wood, the rustling of termites, the trundling steps of millipedes. Butterflies with translucent wings flutter and fight over a bead of sweat. Relieve yourself in the forest and observe the race for the spoils: metallic flies and stingless bees, fifty or more species of scarab beetles in precise sequence, converge in an intense struggle that within hours disperses your human waste into the food chain. Scat reduced to shadow. Imagine what happens to a corpse.

The Indians know, just as they know that the trees sway out of time in order to snap the grip of clinging lianas and slough off great sheets of bark to rid themselves of epiphytes competing for the light. Watching the forest for signs, they anticipate the flowering and fruiting cycles of plants, recognize the preferred foods of animals, exploit the healing power of leaves. Their principal food is derived from the flesh of a toxic root. Poisons from plants enable them to fish and hunt. Their children can track the flight of birds, follow the footsteps of jungle cats. They know, too, of bats that draw blood at night, of fish that enter the urethra and with their spines become lodged in the organs of men and boys, a pain that can scarcely be imagined.

In murky sloughs dwell stingrays and piranha. By night, the riverbanks gleam with caiman. Amidst the reeds and thickets lurk bushmasters, snakes as thick as your arm, all burnished copper with dark markings and the sinister beauty of a creature known to hunt children. In this world of rivers and forest, where anaconda swallow deer and sleep it off in the shallows, peoples of a hundred origins have made a home. Among them are three societies that I came to know: the Bora, botanical wizards of the Río Ampiyacu in Peru, a resurgent culture recovering still from the unimaginable atrocities of the rubber era; the Barasana, descendants of the primordial Anaconda, inhabitants of the most remote reaches of the Colombian Amazon; and the Winikina-Warao, canoe builders of the Orinoco, the largest indigenous society to have survived in Venezuela. Like all of the lowland peoples encountered during my travels, each had a story to tell.

~

ON AN EXPOSED RIVERBANK in the flood forests of the Orinoco delta, I once watched a young woman of the Winikina-Warao dig to the bottom of the world. Her name was Lucia. It only took a few minutes, and she came back with her prize, raw clay to be molded into a fire pad that would allow her to cook for her family on the wood floor of a thatch house that hovered above the ebb and flow of brackish Caribbean tides. A ground fire was out of the question, for the homeland of the Warao is inundated much of the time, and the only permanently dry land runs along distant levees that mark the banks of ancient channels, now overgrown with moriche and manaca, temiche, and other wild palms that sustain the people.

The Warao once lived exclusively in the *morichal*, the palm swamps, away from the rivers, having been displaced from the floodplain thousands of years ago by the Caribs, fearsome cannibals, allies of the werejaguar and the grim lord of the underworld, who swept down the Orinoco and out to the islands beyond the Serpent Mouth of the river. With the arrival of Europeans came disease, which the Warao viewed as the revenge of the spirits, pathogenic arrows unleashed by the gods. Hidden away in a land no one wanted, behind a seawall of red mangrove, the Warao survived the onslaught of El Dorado and remain one of the largest indigenous groups in South America. Today, they live for the most part along the banks of backwater channels, in riverine communities accessible only by boat.

Winikina, where I visited, was a typical settlement, little more than a string of stilt houses without walls, each with its own landing, all linked by a wooden walkway running the length of the village. In the first hours after arriving, I had felt somewhat trapped, confined to a platform with three generations of Warao, a noisy parrot, several monkeys, and a dog, with the river on one side, the flooded ground below, the route to the outhouse a slippery log laid upon the swamp, the outhouse the swamp itself. With me was a good friend, Werner Wilbert, a Venezuelan anthropologist who had lived in the village for well over a year, in the very hut in which we were staying. His father, Johannes, an extraordinary scholar, had studied among the Warao for more than three decades. Both described the delta as the most beautiful world they had known. They were right, but it took several days for me to adjust and understand.

In part, it was a simple matter of the Warao becoming used to our presence, even as we adapted to the transparent privacy that always exists in villages where people live closely together. With Werner's prodding, I surrendered to the reality of the place, dropping any expectation, for example, that I could walk anywhere, anytime. Our routine literally flowed with the river. By morning, we paddled upstream with the tide, filled the dugout with specimens gathered in the flood forest, then drifted back amidst floating islands of purple hyacinth to the village, where we joined the children on the landing as they frolicked and swam in the stream. Every evening, we sat over the river and watched as clouds billowed in the red sky. It felt like being at sea, though the dark brow of the forest rose just behind us.

One day, we visited the grave of Don Antonio Lorenzano, a shaman with whom Werner's father had worked for many years. He was buried, like all Warao, in a protective canoe, which rested on a raised platform, sheltered by palm thatch and surrounded by the forest. It was from this man that Johannes had learned of the role of tobacco in healing ritual, and how the shaman, hyperventilating smoke from enormous ceremonial cigars, brings himself to the edge of nicotine narcosis in his quest for visions and inspiration. Also from Don Antonio had come knowledge of the Lords of the Rain, the House of the Swallow-Tailed Kite, the heraldic raptor, and the dancing jaguar. In a lifetime of study, Antonio had taught Johannes that realms of the spirit could never be extracted from the mundane, that the world of faith and belief was ever constant, like a companion whose shadowy presence is felt even in the darkest night, the most remote backwater, where fish are ever waiting. There was no separation between the spirit and the crude proximity of everyday life. The physical landscape, the material objects of culture, the power of the wind, and the lightness of the clouds were all part of the mystic endowment of the Warao.

"Antonio," Werner explained simply, "revealed to my father the world beneath the surface of things."

That world, according to Johannes, begins with the land itself. In the delta, topography is measured by the inch. The Warao know nothing of mountains, save for the remnants of two petrified world trees, the abodes of earth gods that mark the northern and southern extremes of the universe. One, the Father of Waves, a hillock less than 650 feet high, is located in Trinidad. The other, Karoshimo, rises to less than 500 feet in the piedmont well to the south of the delta. Few Warao have visited these sacred sites. They have never climbed a hill or felt the ragged edge of granite. There are no stones in their homeland. All their perceptions occur at sea level, and the horizon is but a narrow band of dark earth, a sliver separating the black surging waters of the Orinoco from the limitless sky.

The Warao view the Earth as a disk scarred by rivers and floating in a sea. Water saturates everything, and the Earth floats only because it is supported by a serpentine monster whose four-horned head points to the cardinal directions. Anyone who doubts that the Earth disk is delicate and thin needs but to dig a hole or watch as the water seeps into the ground at the base of a tree uprooted by the wind. The disk itself is made of clay, and thus, in order to construct a hearth as Lucia had been doing or to make ceremonial ceramics as her mother might have done in her youth before the traders and missionaries came, the Warao must scrape away a small fraction of the foundation of the world.

From Antonio's teachings, Johannes came to see that symbols of the metaphysical realm are ubiquitous in Warao life, inseparable from the physical experience of the people and inextricably enmeshed in the fiber of their being. The shamanic view of the world is manifest in every aspect of the material culture, not merely in a symbolic sense but because the very construction of the object demands mindfulness and fidelity to ancient religious and customary laws. A sacred rattle, the Calabash of Ruffled Feathers, both protects the community and serves as the *axis mundi* that allows the shaman in death to ascend to a celestial abode, where from the zenith, he travels as a shooting star, a bolt of lightning, or a comet to the house of his patron deity, one of the gods of the north, east or south. A musical instrument, a simple fiddle, provides the music for the dancers, whose grace in movement recalls the sweeping winds of change that have blown so fiercely over the delta in the history of the Warao. Everything thus resonates with the possibilities of another domain, another point in time, another dimension.

The most sacred object of all, Werner explained, was also the most utilitarian, the very canoes that had carried us for days now into the forests. Warao means "owners of canoes," and in a world of water, the people not only travel by canoe, they virtually live in them: sleeping, playing, cooking, trading. To be a builder of canoes is to become a man. Not to possess a canoe is to be relegated among the undistinguished souls of the dead, impoverished, unfed, the lowest of the low. An infant's first canoe is the flat root of a sangrito tree, a plank laid down on the floor of the hut, a surface to practice upon. Before a child can walk, he can paddle, and after a week at Winikina, I grew used to the sight of three-year-old boys and girls, alone, fearlessly maneuvering small dugouts across the wide expanse of the river.

Canoes, in addition to providing an essential means of transportation, moving goods and people throughout the delta, are, more profoundly, the vessels of Warao culture. The toy-like dugout of the child, the discarded hull slowly rotting beneath the landing, the massive seagoing craft that once journeyed to Trinidad and beyond—all represent the mystical knowledge transmitted by the master builder and acquired by the apprentice during their construction, every step of which is dominated by shamanic insight and regulation.

No tree can be felled without the permission of the ancients, the ancestral carpenters who receive offerings of sago starch and tobacco. The spirit of the trees lives on in the canoes, which are carved from the embodiment of Dauarani, the Mother of the Forest, whose womb is both birth canal and coffin. The master builder, who must abstain from sex with his wife until the canoe is consecrated, is visited daily by the spirit of the tree; and as the canoe takes shape as the vulva of the goddess, the very act of carving becomes a mystical act of love, intercourse with the divine.

Seeking what lies beneath the surface of things, as Werner had put it, was the metaphor that had inspired all of his father's life. I had managed over the course of days at Winikina to overcome my ignorance and to sense the beauty and completeness of life in the delta, but Johannes Wilbert had over the course of decades completed the incomparably more arduous task of seeing through to the very essence of Warao identity. The more he learned, the less he knew, or so he would say. What he uncovered beneath the veneer of quotidian life in the delta was an invaluable treasure, a profound insight into another way of being. As a young anthropologist, Werner had the choice of working anywhere he pleased. He chose the Warao, because there was so much more to discover.

ILLUMINATING A WORLD within a world, detecting the underlying symbols of a culture and making sense of what they mean, is a visionary gift of great ethnographers like Johannes Wilbert and Gerardo Reichel-Dolmatoff. I once spent a weekend with Johannes in his cabin high above the Los Angeles hills and for three days listened as he outlined with quiet intensity a totally new paradigm for ethnobotanical research, a way of thinking about plants, people, and landscapes that would allow a true understanding of the relationship between a society and the soil from whence it came, between the cult of the seed and the power of the hunt, the poetry of the shaman and the prose of the priesthood. I left those sessions exhausted yet dazzled.

Reichel-Dolmatoff and Wilbert were good friends, and Reichel-Dolmatoff often came to visit him. Together, they would walk in the botanical garden at UCLA, sometimes joined by Claude Lévi-Strauss, who would fly over from Paris. The thought is beguiling: these three eminent scholars, veterans of a thousand strange cultural encounters, sitting together on a park bench, their imaginations sweeping over the Americas, comparing notes, sharing insights, making plans.

I had met Reichel-Dolmatoff once, but it was a fleeting encounter in the early days of the coca research, and I knew him only through his writings. Each of his many publications is a celebration of wonder. As travelers turn to guidebooks to negotiate the labyrinth of a new city or region, I depended on Reichel-Dolmatoff to reveal the deeper rhythms of a culture, the ebb and flow of nuance and gesture, the actual pulse and essence of the invisible forces encountered while moving through new lands and across unknown frontiers of the spirit. His monograph on the Kogi had been our lens in the Sierra Nevada, but the book I best remember is *Desana: Simbolismo de los Indios Tukano del Vaupés* (published in English as *Amazonian Cosmos*). In the spring of 1975, on the eve of my first visit to the forests of the Vaupés, I was given the book by an old colleague of Richard Evans Schultes, who put me up in Villavicencio, the lowland Colombian town that serves as a gateway to the Northwest Amazon.

Within days of reading about spiritual battles fought by shaman perched on hexagonal shields, all encased in quartz crystals which were themselves the generative organs of Father Sun, I found myself in comparatively mundane circumstances, lost in the Amazon forest not a mile from the Barasana longhouse where I was staying on the banks of the Río Piraparaná near the old Catholic mission of San Miguel. With me, and equally disoriented, was the headman of the village, Rufino Vendaño, who was my guide. Rufino had lived in the immediate vicinity of the longhouse, or *maloca*, all of his life, and to watch him for even a few hours struggle to find his bearings was a revelation. Though there was no panic, his eyes revealed true concern, and when late in the day we stumbled upon a game trail that led us back to the river, he was visibly relieved. He clearly had no more desire to spend a night in the forest than I did.

That evening, at the men's circle, he recounted our misadventure in a charming, self-deprecating manner, which prompted a flurry of similar tales from the Tatuyo and other Barasana living in the maloca. Beneath the easy laughter, however, there was an edge of fear. The forest rose on all sides, the light of a half moon filtered through the thatch, and from the darkness came the sounds of cicadas and tree frogs, the piercing note of a screech owl, the caw-caw-caw of bamboo rats. The clearing around the maloca was the size of a village square, but beyond was a river that flowed to the Amazon, and a forest that stretched to the Atlantic. To reach this place, I had crossed the Andes by truck to Villavicencio, flown three hours in a military transport, and then hired a missionary plane, which had soared into the clouds and burst over the canopy like a wasp, minuscule and insignificant. The forest below was endless, and there was nothing on a human scale. The mission of San Miguel, broken down and long abandoned, was but a minor tear in a formidable tapestry of life. The sense of isolation could not have been more complete.

~

THE LOWLAND FOREST, with its thousand shades of green, envelops and consumes the imagination, and it is only when they are on the rivers that the Indians are able to see the sky. The waterways are not just routes of communication; they are, for the Barasana, the veins of the Earth, the link between the living and the dead, the paths along which the ancestors traveled at the beginning of time. In an astonishing manner, as Reichel-Dolmatoff realized, myth and reality come together in adaptation, a fusion of the past and present that allows the Barasana to cope with the fragility of their lives and thus thrive in an environment that might otherwise so easily overwhelm. Like many nuances of culture, this is not something that the Barasana discuss or even think about. Rather, it is a theme embedded in their very essence, an impulse that lingers along the boundaries of their collective subconscious.

Their origin myth speaks of a great journey from the east, in sacred canoes brought up the Milk River by enormous anacondas. Within the canoes were the first people, together with the three most important plants—coca, manioc, and *ayahuasca*, gifts of Father Sun. On the heads of the anacondas were blinding lights, and in the canoes sat the Mythical Heroes in hierarchical order: chiefs, wisdom keepers, warriors, shamans, and, finally, in the stern, servants. All were brothers, children of the sun. When the snakes reached the center of the world, they lay over the land, outstretched as rivers, their powerful heads forming river mouths, their tails winding away to remote headwaters, the ripples in their skin giving rise to rapids and waterfalls.

Each river welcomed a different canoe, and in each drainage the Mythical Heroes disembarked and settled, with the lowly servants heading upstream and the chiefs occupying the mouth. Thus, the rivers of the Vaupés were created and populated by different peoples. In time, the hierarchy of mythical times broke down, and on each of the rivers the descendants of those who had journeyed in the same sacred canoe came to live together. Still, they recognized each other as family, speakers of the same language, and to ensure that no brother married a sister, they invented strict rules. To avoid incest, a man had to choose a bride who spoke a different language.

When a young woman marries, she moves to the longhouse of her husband. Their children will be raised in the language of the father, but naturally will learn their mother's tongue. Their mother, meanwhile, will be working with their aunts, the wives of their father's brothers. But each of these women may come from a different linguistic group. In a single settlement, as many as a dozen languages may be spoken, and it is quite common for an individual to be fluent in as many as five. Through time, there has been virtually no corrosion of the integrity of each language. Words are never interspersed or pidginized. Nor is a language violated by those attempting to pick it up. To learn, one listens without speaking until the language is mastered.

One inevitable consequence of this unusual marriage rule—what anthropologists call linguistic exogamy—is a certain tension in the lives of the people. The tradition prevents the people of any one river from becoming inbred. With the quest for potential marriage partners ongoing, and the distances between neighboring language groups considerable, cultural mechanisms must exist to ensure that eligible young men and women come together on a regular basis. Thus, the importance of the gatherings and great festivals that mark the seasons of the year. Through sacred dance, the recitation of myth, and the sharing of coca and ayahuasca, these celebrations promote the spirit of reciprocity and exchange on which the entire social system depends, even as they link through ritual the living with the mythical ancestors and the beginning of time. Myth and language, trade and procreation, all are entangled in the challenge of adaptation.

~

As Rufino told our story, we both took a great deal of coca, perhaps too much, for even as the last embers of the fire faded away, I lay awake in my hammock, unable to sleep. Everyone else had long since retired, and in the absence of voices, the maloca came strangely alive. The interior was vast, perhaps one hundred feet long and sixty feet across, with a vaulted ceiling rising thirty feet above the dirt floor. The symmetry of the structure was exquisite:

eight vertical posts spaced evenly in two rows, with two smaller pairs near the doors, crossbeams, and a roof of pleated rows of thatch woven together over a grid of rafters. Still, even to me in a somewhat heightened state of attentiveness, it remained a building, curious and exotic, but a building nevertheless.

Reichel-Dolmatoff experienced it through different eyes. He saw the longhouse as both the womb of the culture and a model of the Barasana cosmos. The roof is the sky, the house beams are the stone pillars and mountains that support it. The mountains, in turn, are the petrified remains of ancestral beings, the Mythical Heroes who created the world. Smaller posts near the doors represent the descendants of the original Anaconda. Overhead, the long ridge pole represents the path of the sun that separates the living from the limits of the universe.

The floor is the earth, and beneath it runs the River of the Underworld, the destiny of the dead. The Barasana bury their dead underneath the maloca, in coffins made from broken canoes, and, going about their daily lives, they walk above the physical remains of their ancestors. To facilitate the departure of the spirits of the dead, the maloca is always built close to water along an east-west axis, since all rivers, including the River of the Underworld, are believed to run east. The placement of the maloca adjacent to a running stream symbolically acknowledges the cycle of life and death, for the water recalls the primordial act of creation in the journey of the Anaconda and Mythical Heroes, and foreshadows the inevitable moment of decay and rebirth.

Outside the longhouse is a world apart, the place of nature and disarray. The owner of the forest is the jaguar, and the demon spirits long ago transformed into animals that eat without thought and copulate without restraint. White people are like the animals, reproducing with such abandon that their numbers swell, spilling over into lands reserved from the beginning of time for the Barasana and the other peoples of the Anaconda. The wild is a place of danger, the origin of disease and sorcery, the realm where shaman go in dreams and where hunters walk each time they leave the protective confines of the maloca and surrounding gardens. When Rufino hesitated in the forest, he had reason to be afraid.

IN A CURIOUS SENSE, the deeper I delved into the esoteric realm of myth and religion, the closer I came to understanding both the raw challenges that confront Amazonian Indians on a daily basis and the cultural mechanisms that allow them to overcome adversity and thrive in a forest homeland that is anything but benign. There is life on the material plane, scarlet macaws sweeping over the canopy at dusk, a field of manioc to be harvested, sweat bees buzzing about at noon. And there is the realm of the spirit, the place where jaguar go and lightning is waiting to be born. The two domains are never confused, nor are they kept apart. The mediator is the shaman, and it is his ability to slip between spheres that allows for the maintenance of the sacred balance, the harmony of social, religious, and political life.

To understand the role of the shaman, and to know anything of his genius in using plants, one must be prepared to accept the possibility that when he tells of moving into realms of the spirit, he is not speaking in metaphor. This was perhaps the most difficult lesson for me to learn as an ethnobotanist schooled in science. But once embraced, it offered a perfectly plausible explanation of how the Indians discovered their useful plants, and thus suggested a possible solution to one of the great mysteries of ethnobotany.

One day in the fall of 1981, during the low point of an expedition to the Peruvian Amazon, I found myself sitting by a fetid slough in the flat light of noon, with the monotonous face of the forest rising on all sides. With me was Terence McKenna, a minstrel of the mystic, who after his death some twenty years later would be eulogized by the *New York Times* as a man who combined a leprechaun's wit with a poet's sensibility to become the Timothy Leary of the 1990s. In the moment, nothing so grand was on his horizon.

"Anyone who says they like the Amazon," he said, "is either a liar or they've never been here. I always feel like a crystal of sugar on the tongue of the beast, impatiently awaiting dissolution."

At the time, Terence and his brother Dennis and I were stranded in the lowlands in a Bora village in the upper reaches of the Río Ampiyacu, the River of Poisons. For ten days, we had been pursuing a curious mystery, the botanical origins of a sacred hallucinogen made from the blood-red resin of several species of virola, trees of the nutmeg family. The basis of a ritual snuff employed by tribes throughout the upper Orinoco, the resins contain a series of powerfully psychoactive compounds, known as tryptamines, that induce not the distortion of reality but rather its dissolution. In fact, they can scarcely be called hallucinogenic, because by the time the effects come on, there is no one home anymore to experience the hallucinations.

Potent as they may be, tryptamines are orally inactive due to the activity of an enzyme, monoamine oxidase, found in the human gut. They may be smoked, injected, or taken through the nose, but not eaten. Yet, alone amongst all the tribes of the Amazon, the Bora and their neighbors the Witoto prepared an orally administered paste, which, according to ethnographic reports, allowed them to commune with the spirits of the forest. Identifying the botanical ingredients and understanding how the preparation worked was the phytochemical challenge that had brought the McKenna brothers to Peru.

By day, we moved through the forest, collecting specimens and bark samples of the various virolas, observing the elders prepare the drug, carefully setting aside for later analysis each stage of the elaboration. In the late afternoon, Terence and I would watch as Dennis lay back in his hammock and swallowed a dose of the latest batch. After more than a week of this, with little to report but nausea and headaches, certainly no little people dancing upon leaves as the shaman had predicted, we had all grown somewhat frustrated.

At night, however, once the experiments were complete, every difficulty was forgotten. The forest itself was transformed. By day, the land lies under a weight of tedium, with vegetation of interest only to the botanist, and the air redolent with fermentation. Then, toward dusk, everything shifts. The air cools, the light softens, and shapes emerge from the forest: flocks of cackling parrots, sungrebes and nunbirds, and in the branches of scandent trees, monkeys and sloths. There is movement in the water, caiman and perhaps an anaconda, its wide head hovering like a periscope. Suddenly, the forest and the water come alive, and the sense of isolation is shattered.

Every evening, we set aside our work and sat for long hours in the men's circle in the longhouse, taking coca and listening to the Bora discuss their day. Often, infused with the plant, we returned to our lodgings by the river, where we spoke long into the night, comparing notes from our travels. I was interested in the role of psychoactive plants in religion and in the healing art of the shaman. For Dennis, it was the shaman's alchemy, the phytochemical mystery itself, that held his imagination. Terence was drawn to the metaphysics, the philosophical implications of plants capable of inducing effects so unearthly, visions so startling, that they had acquired a sacred place in indigenous cultures throughout the world.

I had just come from a month in the northern mountains of Peru, where in the valley of Huancabamba, I had lived with an old *curandero*, a master of a healing cult, the origins of which may be traced in direct lineage to the very

A Yagua shaman, northwest Amazon of Peru, 1983

The grave of Don Antonio Lorenzano,
a Warao shaman, at Winikina, Orinoco Delta,
Venezuela, 1997

The flood forests of the Río Japura,
Brazilian Amazon, 1982

dawn of Andean civilization. As I explained to Terence and Dennis, even today people from all over South America go there to seek guidance and treatment for a plethora of ailments. Once or twice a week, every week, a new set of acolytes assembles in the courtyard of the curandero's farmstead and patiently awaits the darkness.

The ceremony is complex, lasting well into the night, and everyone in the healing circle imbibes through the nostril as much as half a quart of raw cane alcohol infused with tobacco or datura leaves. At midnight, the curandero dispenses a decoction prepared from San Pedro, the Cactus of the Four Winds. During the ensuing mescaline intoxication, the curandero diagnoses each patient's ailment. But treatment can take place only the following day, at the end of a long pilgrimage that carries the patients far higher into the mountains to a series of sacred lakes, around whose periphery grow the medicinal plants that are alone believed to be therapeutic. Before the herbs are administered, the patients participate in a lengthy ritual, culminating in a baptismal plunge into the frigid water of the lakes.

The metaphor is clear. In order to heal the body, aspirants must seek spiritual realignment through the use of the magic plant, as well as move through geography, enduring physical hardships to reach the sacred lakes, where only after a ceremony of metamorphosis can they be open to the pharmacological possibilities inherent in the medicinal plants. Here, I suggested to my friends, was the essence of the shamanic art of healing: a fusion of mind and spirit, plant and landscape, sacrifice and yearning.

In the West, the shaman is often regarded as a harmless figure, a gentle elder who heals with feathers and beads and incantations. In my experience, however, most shaman are just a little crazy. That, after all, is the nature of the calling. The shaman, as Joseph Campbell said, is the one who swims in the mystic waters the rest of us would drown in. Indeed, he or she chooses to enter realms that most people do not want to even imagine.

Most indigenous people are as happy to relegate affairs of the spirit to the shaman as we are to leave such issues to our priests. In Western society, however, we make a distinction between religion and medicine, whereas in most indigenous traditions, priest and physician are one, for the state of the spirit determines the state of the body. Thus, to treat disease, to address the cause of misfortune, the shaman must invoke some technique of ecstasy that allows him to soar away on the wings of trance to reach those distant metaphysical realms where he can work his deeds of medical and spiritual rescue. Hence the use of psychoactive plants.

The pharmacological effects of these preparations are stunning, and their place in shamanic life revelatory. But what interested Dennis as a phytochemist was the mystery of their origins. He spoke at length about ayahuasca, the vine of the soul, a sacred brew that has fascinated travelers in the Amazon since first being reported by the English botanist Richard Spruce in the mid-nineteenth century. I knew the plant, and at Schultes's behest had sampled the potion on several occasions. But Dennis was one of the world authorities, and as he spoke long into the night, I truly came to understand for the first time the genius of the shaman and the allure of the mystery that had attracted Dennis and Terence to this forgotten village on the banks of a little known affluent of the Rivers Amazon.

Ayahuasca, also known as *yagé* or *caapi*, is a preparation derived from two species of Amazonian lianas, *Banisteriopsis inebrians* and, more commonly, *Banisteriopsis caapi*. The potion is made in various ways, but long ago, the shaman of the Northwest Amazon discovered how to enhance the effects by adding a number of other plants. With the dexterity of a modern chemist, they recognized that different chemical compounds in relatively small concentrations may effectively potentiate one another. In the case of ayahuasca, some twenty-one admixtures have been identified, including most notably *Psychotria viridis*, a shrub in the coffee family, and *Diplopterys cabrerana*, a forest liana closely resembling ayahuasca. Unlike ayahuasca, both of these plants contain tryptamines.

"The only way a tryptamine can be taken orally," Dennis explained, "is if it is taken with something that inhibits monoamine oxidase, the enzyme in the stomach. Amazingly enough, the beta-carbolines found in ayahuasca are precisely the kind of inhibitors necessary for the job."

In other words, when the bark of the banisteriopsis liana is combined with either the bark or leaves of these admixtures, the result is a powerful synergistic effect. The visions become brighter, and the blue and purple hues induced by banisteriopsis alone are augmented by the full spectrum of the rainbow.

"Now I ask you," Dennis said, "how on earth did they figure it out? What are odds against finding in a forest of fifty thousand species, two plants, totally different, one a vine, the other a shrub, and then learning to combine them in such a precise way that their unique and highly unusual chemical properties complement each other perfectly to produce this amazing brew that dispatches the shaman to the stars? You tell me."

Many ethnobotanists avoid the question by invoking trial and error, a catchall phrase that explains very little, since the elaboration of the preparations often involves procedures that are exceedingly complex or that yield products of little or no obvious value. An infusion of the bark of *Banisteriopsis caapi* causes vomiting and severe diarrhea, reactions that would hardly encourage further experimentation. Yet not only did the Indians persist, but they developed dozens of recipes, each yielding potions of various strengths and nuances for specific ceremonial and ritual purposes.

"I don't think there is a scientific explanation," Terence remarked. "And if there is, why should it take precedence over what the Indians themselves believe? They say they learn in visions, that the plants speak to them. They're not making it up to please us. It's what they have always believed."

The Indians have their own explanations, of course: rich cosmological accounts of sacred plants that journeyed up the Milk River in the belly of the Anaconda, potions created by the primordial jaguar, the drifting souls of shaman dead from the beginning of time. As scientists, Dennis and I had been taught not to take these myths literally. Terence, who suffered from no such constraints, suggested that their botanical knowledge could not be separated from their metaphysics.

"Have you ever been in the upper Putumayo with the Ingano or the Siona?" I asked, referring to tribes with whom Schultes had lived in the 1940s in the lowlands of Colombia. The Ingano, I explained, recognize seven varieties of ayahuasca. The Siona have eighteen, which they distinguish on the basis of the strength and colors of the visions, the authority and lineage of the shaman, even the tone and key of the incantations that the plants sing when taken on the night of a full moon. None of these criteria makes sense scientifically, and, to a botanist, all the plants belong to a single species, *Banisteriopsis caapi*. Yet the Indians can readily distinguish the varieties on sight, and individuals from different tribes can identify these same varieties with remarkable consistency.

"Imagine what it means," Terence said, "to really believe that the plants sing to you in a different key, to have a taxonomic system that is consistent and true, based on an actual dialogue with the plants."

In the end, Dennis did manage to solve the enigma of the Bora pastes. It turned out that the resins themselves contained, in addition to various tryptamines, other compounds in small concentrations that served to inhibit monoamine oxidase and thus potentiate the drug. But, curiously, he was never able to experience the effects himself, at least nothing close to the intensity of the visionary intoxications reported by the Bora and Witoto shaman. His research earned him a doctorate and was widely heralded, but in his own mind a large part of the mystery remains.

Though acknowledged as the greatest Amazonian plant explorer of his generation, Schultes always claimed to rank as a novice in the company of shaman. Like so many of his acolytes, Dennis and I had been drawn to the Amazon to seek its gifts: leaves that heal, fruits and seeds that provide the foods we eat, plants that could transport the individual to realms beyond reason. But in time we both came to realize that in unveiling indigenous knowledge, our task was not merely to identify new sources of wealth but to understand and celebrate a distinct vision of life itself, a profoundly different way of living in a forest. This is something that Terence always knew.

IT MAY SEEM ODD, given how fortunate I had been in my choice of vocation, but after eight years of thinking only of plants, their place in culture, and the wonder of the Amazon and Andes, I grew restless, eager for change. Not that I regretted the months of fieldwork, the thousands of specimens collected, but my work had reached a certain plateau.

After three years of research, the story of coca was essentially known. As an ethnobotanist, I had surveyed other plants as well and encountered a number of unusual mysteries. In northern Peru, a casual collection yielded a new psychotropic cactus. The valley of the moon in Bolivia revealed yet another hallucinogen, closely related to the Cactus of the Four Winds, *Trichocereus pachanoi*, the sacred plant that had sparked the rise of civilization in the Andes two thousand years before the birth of Christ. The pursuit of an admixture used with coca led to one of the first descents of the Apurimac and later the Urubamba, headwaters of the Amazon. For nearly a decade, my every thought had been a plan to return to one of the lowland societies I had come to know: in Bolivia, the Chimane, Mosetene, and Tacana; in Peru, the Machiguenga, Shipibo, Bora, and Yagua; in Ecuador, the Shuar, Kofán, Siona-Secoya, Waorani, and Quichua. From Colombia beckoned the Kamsa and Ingano, Embera, Barasana, Witoto, Tukano, Cubeo, Makuna, Tikuna, and a host of other peoples who, true to their essential spirit, had always welcomed me, an itinerant scholar, as they would any sympathetic outsider.

But I was tired of simply documenting plants recognized by a particular culture. Once compiled, these academic reports seemed to me little more than grocery lists, devoid of scientific content. I wanted to use plants, and the genius of people who manipulated them, to ask larger questions. In part, this impulse was driven by intellectual aspirations, simple curiosity, really, but it also reflected a certain impatience, a reflexive tendency to move on just as things were becoming comfortable. The more experience I had as a plant explorer, the more I yearned for something completely novel. Fortunately, I was based at the right institution, for a new challenge always loomed at Harvard's Botanical Museum, in the fourth-floor aerie of Professor Schultes.

One morning, as I was attempting to explain to an undergraduate student the nuances of Xavante kinship, word came that the professor wanted to see me. As soon as I could, I abandoned my young charge, raced up the iron steps of the museum, and burst into his office only to encounter the university president, Derek Bok. Spewing apologies, I retreated for the door, but was stopped by Schultes, who politely asked Bok to step outside for a moment as he had a student to see him. As the president of Harvard shuffled out, a smile on his face, I took his place in a chair across from the great professor's desk. At this point, I would have done anything for him. So when he asked if I might be interested in traveling to Haiti to seek the formula of a powder used to make zombies, I didn't hesitate. I accepted the assignment, not knowing that it would, in the end, consume four years of my life.

The Río Ampiyacu, River of Poisons, Peruvian
Amazon, 1981

A Barasana boy with a scarlet macaw, Río Piraparaná, northwest Amazon of Colombia, 1977

Caboclo mothers and children, Amazon Delta, Brazil, 1987

Vodoun acolytes possessed by Damballa-Wedo,
the serpent god, Saut d'Eau, Haiti, 1982

CHAPTER FOUR

THE FACE OF THE GODS

On Easter Sunday, 1982, I returned from Haiti to New York and was strolling through the customs hall at John F. Kennedy Airport when, suddenly, I was accosted by a nearly hysterical agent of the U.S. Department of Agriculture. Word had reached her that an anthropologist, with some knowledge of Vodoun, had just passed through immigration.

I expected her to ask for permits. In my suitcase were dried toads and snakes, seeds and herbs, powders made from toxic fish, a small collection of amulets, fetish symbols, and bits and pieces of human bones destined for chemical analysis. Ignoring my luggage, she reached her hand into the cleavage of an ample bosom and withdrew a large silver crucifix. With breathless urgency, she asked, "Will this help?"

"Excuse me?" I replied.

"The Haitians!" she roared. "The Haitians!"

Only then did I realize that the fate of American agriculture was the last thing on her mind. What she wanted to know was whether the cross would protect her from the Haitian immigrants, voodoo and devil worshipers all, as she put it, then entering the country. I told her not to worry and walked on. In my backpack was a live toad, six inches across, with enough venom in its glands to kill half a dozen people.

My visit to Haiti had been prompted by an unusual assignment. A team of physicians and scientists, led by Nathan Kline, a pioneer in the field of psychopharmacology, had discovered the first documented instance of zombification. What made the case unique was the fact that the putative victim, a middle-aged man named Clairvius Narcisse, had been pronounced dead at an American-directed hospital that kept impeccable records. The demise of Narcisse had been documented by two physicians, both American trained, one an American, and witnessed by a sister of the deceased. Many years after the burial, to the horror and astonishment of family members, Narcisse had wandered back into his native village, where he presented a chilling tale of having been victimized by a sorcerer and transformed into a zombie. Lamarque Douyon, Haiti's leading psychiatrist, conducted a thorough investigation and, in collaboration with Nathan Kline, went public in 1980 with the stunning conclusion that Narcisse's account was true.

A zombie, according to folk belief, is the living dead, an individual killed by sorcery, magically resuscitated in the grave and exhumed to face an uncertain destiny, a fate invariably said to be marked by enslavement. Kline and Douyon, of course, did not believe in magic, and knew there had to be a scientific explanation. They focused their attention on reports of a folk preparation, mentioned frequently in the popular and ethnographic literature, which was said to induce a state of apparent death so profound as to fool a Western-trained physician. The Haitian people evidently accepted the existence of the poison, for it is specifically mentioned in the penal code of the country and penalties for its use are severe. Since the medical potential of a drug capable of inducing apparent death could be considerable, especially in the field of anesthesiology, Kline, after contacting Schultes at Harvard, dispatched me to Haiti in the spring of 1982 with the goal of securing the formula of the preparation.

FROM THE START, I pursued two avenues of research. To satisfy Kline and his colleagues, I sought to document the elaboration of the poison, in the hope of identifying a natural substance that could actually induce apparent death. But the higher goal was to make sense out of sensation, to provide a context for a folk belief that had been exploited in a lurid, even racist, manner to discredit an entire people, their culture, and their religion. Though sent to Haiti to seek the chemical basis of a social phenomenon, I would, in the end, explore instead the psychological, spiritual, political, and cultural dimensions of a chemical possibility. As it turned out, the zombie phenomenon was one small dark thread woven through the rich and colorful fabric of the Vodoun worldview.

When I arrived in Haiti, I knew little of the culture. But it did not take long to realize that a chasm existed between reality and the popular misconceptions that had sent the agricultural agent at Kennedy Airport into a paroxysm of fear and anxiety. The capital of Port-au-Prince was a sprawling muddle of a city, but life in the streets had a rakish charm. Street-side vendors in alleys damp with laundry hustled their herbs, market women sauntered with exquisite grace along broken boulevards while down by the docks where the cruise ships glittered, men with legs as hard as anvils dragged carts piled high with bloody hides. Children were everywhere, their angelic faces laughing. As I drove into the city, a solitary figure, dressed in white, quite sane and perfectly harmless, stepped into the road, halting all traffic as he stood alone, dancing with his shadow.

That evening, I visited Max Beauvoir, a Vodoun priest, or *houngan*, whose name had been given to me by Nathan Kline. An organic chemist by training, Beauvoir had worked in the United States and France, but had found his calling in Vodoun upon his return to Haiti, in a moment of revelation inspired by the death of his grandfather. A man of great charm and generosity, fluent in several languages, Max would over the course of many months lay his country before me as a gift.

He began that very night with an invitation to a ceremony at his *hounfour*, or temple, located south of the city along the Carrefour Road. There, with the wind coming off the sea, I watched in awe as white-robed initiates invoked the spirit. Responding to the rhythm of the drums, the invocations of the priestess, the resonance of songs, they moved as a single body, circling the center post of the temple. The dance was deceptively simple: feet flat to the ground, small shuffling steps, shoulders and hips moving with a fluidity that seemed to have all of nature swaying in sympathy. The voice of the priestess was clear and powerful, slicing through the joy and celebration. Libations were blown to the wind. Small clouds of dust hovered above the dry ground. A continuous battery of sound drove each moment forward.

As the energy rose, the rhythm of the drums unexpectedly shifted to a highly syncopated, broken counterpoint that created a moment of excruciating emptiness. A void enveloped the dancers, who slipped into stillness only to explode in spasmodic pirouettes that carried them, still spinning, into the glow of burning fires. They danced on the red-hot coals with impunity. A young man took an ember the size of a small apple into his mouth, tight between his teeth. His hot breath sent sparks throughout the temple. The ritual was theatrical and strange, impossible to understand except on its own terms.

"White people go to church and speak about God," a Haitian friend later told me, "we dance in the temple and become God." Once possessed, the believer loses all consciousness and sense of self; he or she becomes the spirit, taking on its persona and powers. How can a god be harmed?

The fact that the coals do not burn the believers is an astonishing example of how the mind, unleashed during a state of extreme excitation, can affect the body that bears it. But perhaps more profoundly, it offers a raw demonstration of the power of faith, a visceral experience of the metaphysical. Within hours of arriving in Haiti, I had witnessed a phenomenon that had eluded me in the Amazon for a decade: a window open wide to the mystic.

IF ANY OF US were to be asked to name the great religions of the world, how would we respond? Buddhism, Christianity, Judaism, Islam, Hinduism? What continent would be left out? No doubt sub-Saharan Africa, the tacit assumption being that African peoples had no organized faith. But, of course, they did, for what is religion but a set of ideas and beliefs that satisfies the spiritual needs of a people?

Vodoun is a Fon word from Dahomey meaning simply "spirit" or "god," and the practice of Vodoun is but the distillation of profound religious ideas that came over from Africa during the tragic diaspora of the slavery era. Sown in the fertile soil of the New World, brought into being by dreams of redemption and sacrifice, the spiritual intuitions of the ancient homeland burst forth in a dozen original forms. There is Obeah and Cumina in Jamaica, Candoblé and Macumba in Brazil, Hoodoo in the American South, and, with its rich overlay of Catholicism and medieval magic, Santaria in Puerto Rico, Cuba, and the other Hispanic countries of the Caribbean.

The image of Vodoun as something evil, a black magic cult, grew out of fear rooted in ignorance, reinforced by pernicious clichés. Beginning in 1915, the United States Marine Corps occupied Haiti, and for the next twenty years, or so it seemed, nearly every visitor landed a book contract, unleashing a slew of pulp fiction titles such as *Black Baghdad*, *Cannibal Cousins*, *The White King of La Gonave*, *A Puritan in Voodooland*.

These, in turn, inspired a succession of Hollywood horror movies: *I Walked with a Zombie*, *The Night of the Living Dead*, *Zombies on Broadway*. Replete with images of the macabre, zombies crawling out of the ground, children bred for the cauldron, these otherwise forgettable books and films, produced at the height of the Jim Crow era, essentially suggested to the American public that any country that tolerated such abominations could only find its redemption in military occupation.

In truth, Vodoun is a benign faith that, like all great spiritual traditions, encompasses a complex metaphysical worldview. In many ways, it is the quintessentially democratic religion, for the believer not only has direct access to the spirit realm but actually receives the gods into his or her body and becomes transformed.

Vodoun essentially acknowledges the reciprocal relationship between the living and the dead. Those of this earthly plane must honor their ancestors, who give birth to the gods. The dead, in turn, if properly served, are expected to assist the living. Death is not feared for its finality but is regarded as a crucial and vulnerable moment when the spiritual and physical components of men and women separate.

A proper death is one in which the *ti bon ange*, "the little good angel," the element of the soul that creates personality, character, and willpower, goes beneath the Great Water, only to be ritualistically reclaimed a year and day later by a priest who places the departed spirit in a small clay vessel, which is stored in the inner sanctuary of the temple. In time, this spirit, initially associated with a particular individual, becomes part of the vast ancestral pool of energy, out of which emerge the *loa*, the 401 spirits of the Vodoun pantheon.

Yet, in this remarkably dynamic faith, even the dead must be made to serve the living; and in order to serve the living, they must be invoked by ceremony to become manifest, returning to Earth to displace the soul of the living. For the believer, spirit possession is the touch of divine grace, a moment of transcendence, the epiphany of the Vodoun faith.

The spirits live beneath the Great Water, dividing their time between Haiti and the mythic homeland of Guinée, of Africa. But they often chose to dwell in places of natural beauty. Those who serve the loa are drawn to these sites much as Christians are drawn to cathedrals, not to worship the place or the building, but to be in the presence of God. In the north, there is the festival of Plaine du Nord, where once each year a mud pond spreads over a dry roadbed, and pilgrims gather to fill their bottles with curative potions derived from the earth.

Around the periphery, initiates light candles and worship at the roots of a sacred tree, as men and women slip into the basin and emerge transformed, bodies coated with clay, eyes wide, hands reaching out to feed herbs to cattle waiting for the machete of Ogun, god of war, to strike and spread their blood across the surface of the pool. Young women straddle the dying animals, as boys and fathers wallow in the mud, singing, laughing, playing.

Farther south, there is the waterfall of Saut d'Eau, where Erzulie, goddess of love, escaped the wrath of the Catholic priests by turning into a pigeon and disappearing into the iridescent mist. The cascade is also the domain of Damballah-Wedo, the serpent god, repository of all wisdom and the source of all falling waters. When the first rains fell, a rainbow, Ayida Wedo, was reflected. Damballah embraced Ayida, and their love enveloped them in a cosmic helix from which all creation emerged. Thus, once each year, Vodoun acolytes, ten thousand strong and dressed in white robes, gather at the sacred waterfall. Arriving on foot, they drift across the limestone escarpment with the motion of night clouds and descend by a trail to a basin carved into the rock face.

There, beneath the spreading branches of a sacred mapou tree, illuminated by the glow of a thousand candles, they enter the water. Merely to touch the cold thin blood of the divine, to step behind the veil of the falling water, is to become possessed by Damballah-Wedo. At any moment, there are dozens of pilgrims, possessed by the spirit, slithering across the wet stones like snakes.

In the radiant light of the waterfall, all thoughts of Vodoun as something fearful disperse. There are children laughing, women shamelessly naked, men whose power and devotion defy the force of the falls. One man steps forward in trance, fully clothed, and enters the most thunderous

Estelle Beauvoir at the sea of Mariani, Haiti, 1982

Empowered by the spirit, Vodoun initiates embrace fire with impunity, south of the Carrefour Road, near Port-au-Prince, Haiti, 1982

place in the cataract. The entire weight of the waterfall strikes his body. The water is cold, frigid to a Haitian. He stands like a statue, face to the rocks, as his clothes are literally torn from his body. Stripped bare, he turns, an act of both defiance and surrender. Like a serpent that has shed its skin, he waits patiently for renewal. There are shouts from the children, and houngan and *bokor*, priest and sorcerer, rush to his side to pay homage to his courage. Everything flows away into the promise of another year.

~

IT WAS AGAINST this backdrop of faith, this sacred alignment of the light and the dark, that I set out to examine the phenomenon of the Haitian zombie. The formula of the elusive poison was the key. From preliminary reports, I knew even before leaving for the Caribbean that the preparation had to be topically active, capable of inducing a prolonged psychotic state, and that the initial dosage had to bring on a deathlike stupor. It had to be extremely potent and, because both the toxin and its purported antidote were likely to be organically derived, the sources had to be plants or animals found in Haiti.

Many plant and animal substances can kill, but if the Narcisse case was to be believed, the preparation had to be capable of provoking the misdiagnosis of death, a far more curious outcome. Although each sorcerer had a unique formula—crushed seeds and leaves, spiders, snakes, toads, magical powders, and human bones—the consistent and critical ingredients turned out to be species of marine fish, all belonging to an order known as the Tetraodontiformes, whose viscera and skin contain a nerve poison known as tetrodotoxin.

Among the most poisonous substances known from nature, tetrodotoxin is roughly a thousand times stronger than sodium cyanide: a lethal dose of the pure toxin could rest on the head of a pin. More compelling than the sheer toxicity of the drug is the manner in which it affects the body, causing metabolic rates to fall dramatically low. The pulse becomes imperceptible and peripheral paralysis is total. Though quite unable to move, the victim remains fully conscious until the moment of actual death.

In Japan, several species of fish closely related to those sought by the sorcerers in Haiti have long been eaten as a delicacy. Specially trained chefs, licensed by the government, carefully remove the toxic organs, reducing but not eliminating the poison, so that the connoisseur still enjoys the exhilarating physiological effects of a mild intoxication. For the Japanese, consuming *fugu*, as the fish are known, is the ultimate aesthetic experience. It is also highly dangerous. On reviewing Japanese and Australian medical literature, I was astonished to find case after case of individuals who, having been pronounced dead by physicians, had returned to the realm of the living. In one instance, a man, sealed in a coffin ready for burial, awoke in the darkness to a unique world of terror. Only by chance did a passing attendant hear his screams and come to the rescue. To avoid such a fate, folk tradition in rural areas of Japan dictates that those who succumb to fugu poisoning be laid out beside the grave for three days to make sure that, before burial, they are really dead.

These remarkable cases, although unrelated to events in Haiti, nevertheless proved beyond doubt that the Haitian sorcerers had found a natural substance that, if administered in proper dosage, not only could make someone appear to be dead but had, in fact, done so many times in the past. That Clairvius Narcisse's symptoms were consistent with the known effects of tetrodotoxin suggested at least the possibility that he had been exposed to the poison. If this did not prove that he was a zombie, it did, at least, substantiate his account.

WHILE THE FORMULA of the preparation took the zombie from the phantasmagoric into the realm of the plausible, it by no means solved the essential puzzle. No drug, of course, can create a social phenomenon; it merely provides a template upon which cultural and psychological forces can go to work on the individual. Those Japanese who succumb to fugu do not become zombies; they are poisoning victims. In recovery, they can rationalize the dreadful experience within the constraints of their own cultural and personal expectations.

Clairvius Narcisse, on the other hand, had been raised from birth to believe in the reality of zombies, and he would have carried his fears into the grave and beyond. Until I understood his reality, the matrix of beliefs that cushioned and provoked his imagination, I could not claim to know anything about the zombie phenomenon. For the Vodounist, the formula is but a support of the magical force of the sorcerer, and it is this power, not a poison, that creates the zombie.

It was at Saut d'Eau, as I lay beneath the spreading branches of a mapou tree, that I finally came to understand. The Vodounist believes that there are two kinds of death: those that are natural, acts of God beyond the reach of sorcery; and those that are unnatural, mediated by the bokor. Only those who die an unnatural death may be claimed as a zombie. The poison is an effective tool to induce such a death, but the performance of a magical rite is what actually creates the zombie.

The bokor gains power over the individual by capturing the victim's ti bon ange. A zombie appears cataleptic precisely because it has no soul. Robbed of personality, character, and willpower, the body is but an empty vessel subject to the commands of an alien force, the one who maintains control of the ti bon ange. The notion of external forces taking control of the individual, and thus breaking the sacred cycle of life, death, and rebirth that allows human beings to give rise to the gods, is what so terrifies the Vodounist. This explains why the fear in Haiti is not of zombies but rather of becoming one.

Understanding zombification from the perspective of the believer led to yet another revelation. Though there was no doubt that tetrodotoxin could induce apparent death, and that the fish used by the bokor contained the poison, the preparation itself was obviously not made with the precision of a modern pharmaceutical laboratory. Indeed, levels of tetrodotoxin among individual fish vary greatly, and at certain times of year, as much as half of the population of any given species may contain none of the drug at all. Thus, any particular batch of the powder may be totally inert. But the bokor does not have to account for his failures. If he administers a preparation that has no effect, he can claim that his magic was deflected by the intervention of a benevolent priest. If, on the other hand, the victim actually dies, the bokor can suggest that the death was a call from God and beyond the reach of his sorcery. A bokor's failed attempts do not count, only his successes. Even if the poison was effective but once in dozens of attempts, the outcome would support the powerful reputation earned by the zombie phenomenon. Indeed, for any number of reasons, zombification must be an exceedingly infrequent and highly unusual occurrence. However, the phenomenon's power depends not on how often it occurs but rather on the fact that it can and apparently has occurred.

While the potency of the sorcerer's spell and the powder itself suggest a means by which zombies may be created, now or in the past, they nevertheless explained very little about the place of zombification within Haitian traditional society. The peasant knows that the fate of the zombie is enslavement, a concept that suggests the victim suffers a fate worse than death: the loss of individual freedom implied by slavery, and the sacrifice of individual identity and autonomy implied by the loss of the soul, the ti bon ange. The ability to cast a victim into the purgatory of zombification confers immense power on the bokor, should he chose to exercise it. But how and why is someone chosen to become the victim of the bokor's sorcery? Answering these questions prompted the final and most challenging phase of the investigation.

FROM THE CASE OF NARCISSE and reports of other alleged zombies, it appeared that the threat of zombification was not invoked in a random manner. All of them were pariahs in their communities at the time of their demise. Members of Narcisse's family claimed that he had been brought before a tribunal to be judged. When he returned to his village for the first time, his family did not doubt his identity, but nevertheless ordered him away. His brother, who according to Narcisse had been responsible for his demise, had lived out his natural life in the village. The bokors who administered the powders and wielded the spells also lived within their communities. Whatever their perceived power, it is difficult to imagine any sorcerers being long tolerated unless their activities ultimately served the needs of their communities.

In my travels in Haiti, I had often heard of the Bizango, a secret society feared by many Haitians that constituted a force, if not an institution, parallel to the Vodoun temples headed by the priests. Several bokors had told me that the Bizango controlled the zombie powders. Throughout Equatorial West Africa, the fountain of Haitian culture, secret societies were acknowledged to be the most powerful arbiter of social and political life. The connection between these clandestine groups and the Bizango had been traced in direct lineage by a respected Haitian scholar, Michel Laguerre. In Africa, the societies are known to have a judicial function as tribunals that apply sanctions. To punish those who violate the codes of their communities, they exploit the toxic power of plants.

The link to zombification was evident, and in the last months of my sojourn in Haiti I was able to enter the Bizango and undergo preliminary training as an initiate. What emerged was a clear sense that beyond the ritual activities, the flamboyant rites that strike terror in the hearts of so many Haitians, the secret societies constitute a true and effective political force that protects community resources, particularly land, even as they define the power boundaries of the villages. Sorcery and poisons are their traditional weapons, and within the Bizango, there is a complex judicial process by which those who violate the code of the society may be punished. Zombification is the ultimate sanction. Clairvius Narcisse, it seems, was no innocent victim. According to the terms and structure of his society, his condition had been deserved, his fate sealed, by his own misdeeds.

ALTHOUGH IT WAS ULTIMATELY impossible to prove that Narcisse had received a dose of the poison, or that he had passed close to death and been revived, his case, provocative as it was, provided an extraordinary conduit to culture and obliged the scientific world to take seriously a phenomenon that had historically been dismissed, even ridiculed, often in a racist manner. When the investigation began, even sympathetic students of Vodoun viewed the popular belief in zombies as but a lamentable example of a nation's notorious instincts for the phantasmagoric. By the time it ended, there was no longer any doubt that the bokor had identified a natural product that not only could induce a state of apparent death but had done so many times in the past, as is evident in the medical literature. The purpose of science is not to discern absolute truth but, rather, to generate better ways of thinking about phenomena. Twenty years on, the link between the toxic powder and zombification remains compelling, and the hypothesis still stands.

But when I think back to my time in Haiti, it is not the zombie research that I recall as much as the Vodoun culture itself. For the first time, I was exposed to a belief system so profound in its implications and to a people so distinct in their perceptions that they provoked a fundamental shift in my own worldview. Not that I became an acolyte. Vodoun embodies not only a set of spiritual concepts but prescribes a way of life, and, like most religions, it cannot be abstracted from the day-to-day lives of the believers; spiritual convictions are fused into their very being. From my perspective, I could no more become a Vodounist than I could become a Haitian, and to suggest otherwise was to deny and, in a sense, betray the power and depth of their devotion. But what I could do was bear witness to an amazing culture in which there was no separation between the sacred and the profane, between the material and the spiritual. A land where every dance, every song, every action, was but a particle of the whole, each gesture a prayer for the survival of the entire community.

The first night I was in Port-au-Prince I watched a man in a state of trance carrying in his mouth a burning ember for several minutes. I knew of other societies where believers affirm their faith by exposing themselves to fire. In Brazil, Japanese immigrants celebrate the Buddha's birthday by walking across beds of coals. In Greece, firewalkers believe that they are protected by the presence of Saint Constantine. The same sort of thing goes on in Singapore and throughout the Far East. Western science accounts for these extraordinary feats by invoking the same effect that makes drops of water dance on a skillet. The theory suggests that, just as heat vaporizes the bottom of the water droplet as it approaches the skillet, a thin protective layer of vapor is formed between the burning rocks, for example, and the firewalker's feet.

To my mind, this explanation begged the question entirely. A water droplet on a skillet is not a foot on a red-hot coal, nor teeth grasping an ember. I still burn my tongue if I place the lit end of a cigarette on it. After what I had witnessed in Haiti, any explanation that did not take into account the play of mind and consciousness, belief and faith, seemed desperately hollow. I had no experience or knowledge that would allow me to rationalize or to escape what I had seen. The man had entered the spirit realm with ease and impunity; his body had not been harmed by the fire. Here was proof of the power of conviction.

HAITI LEFT ME CONVINCED that cultural beliefs really do generate different realities, separate and utterly distinct from the one into which I had been born. During subsequent years, as my travels led to other parts of the world, the forests of Borneo and the Tibetan plateau, the deserts of East Africa and the ice floes of the Arctic, memories of Haiti continued to serve as a lens through which I took the measure of a place, knowing always that each culture represented, by definition, a unique facet of the human legacy and promise. The more intensely I embraced this notion, the more I came to see in the sweep of modernity an impending catastrophe, as we drift away from diversity, as languages are lost and ancient peoples are convulsed in an upheaval of violence and transformation.

A Vodoun priest, or houngan, prepares a healing potion, Saint Marc, Haiti, 1984

Facing page: The teachings of the Bizango Shanpwel, Saint Marc, Haiti, 1984

The public face of the Bizango Shanpwel, the secret society, with men dressed as women during the Rara celebrations of Easter, Artibonite Valley, Haiti, 1982

A poison maker, Léogane, Haiti, 1982

The arrival of the spirit at the Peristyle de Mariani, a Vodoun temple on the Carrefour Road near Port-au-Prince, Haiti, 1982

Vodoun acolytes, or hounsis, at the mud basin of
Plaine du Nord, Haiti, 1982

Transformation and metamorphosis,
Plaine du Nord, Haiti, 1982

His clothes torn away by the force of the cataract,
a man emerges renewed, Saut d'Eau, Haiti, 1982

Kowe, a Jaguar shaman of the Waorani, Amazonian lowlands of Ecuador, 1981

CHAPTER FIVE

THE LAST NOMADS

IN A BITTER MOMENT OF RESIGNATION, Claude Lévi-Strauss said that the people for whom the term cultural relativism was invented have rejected it. Indigenous societies throughout the world, he suggested, have deliberately abandoned the old ways and chosen instead to embrace the uncertain promise of the new. In this case, the great anthropologist was only half right. Traditional cultures have survived precisely because of their ability to cope with change, the one constant in history. People disappear only when they are overwhelmed by external forces, when drastic conditions imposed on them from the outside render them incapable of adapting to new possibilities for life. In eastern Ecuador, I lived in a Kofán village that was destroyed in a single generation by the discovery of oil. When I returned later, the shaman I had worked with was dead, and his son had a job with Texaco. In Colombia, Barasana men whom I knew well have been reduced to coca-growing serfs by the drug lords and their allies, the revolutionary guerrillas of the left. Cattle barons tear down the forests of the Bora and Witoto. In Africa, warriors who once fought with spears now have their choice of automatic weapons.

It is not change that threatens the integrity of the ethnosphere, it is power, the crude face of domination. Given a chance, indigenous societies can thrive in a period of flux and transformation. But, as in any time of turmoil, there are risks, and the consequences can be dire, as is evident in the stories of the Waorani and Penan, rain forest peoples of the Amazon and Borneo, and of the Rendille and Ariaal, nomadic herders from the deserts of East Africa.

KOWE WAS A JAGUAR SHAMAN, too old to hunt, but just the right age to empower the poison darts. Younger men made the actual curare, scraping the bark of the liana, placing the shavings into a funnel of palm leaves suspended between two spears, slowly percolating water through, and collecting the drippings in a small clay vessel. The dark fluid was brought to a frothy boil, cooled, and fired again until a thin viscous scum formed on the surface. Then, just as the curare congealed, Kowe would take his place on a wooden stool by the fire, his feet resting on a stone ax, and slowly spin the tip of each dart in the resinous poison. One by one, he placed the darts into the earth, and as the tar hardened into a jet-black lacquer, he sang the forest into being, sliding his voice up and down in a chant that only the Jaguar Mother had the power to translate for the world. His voice rose to the black and smoky rafters of the lodge, mingling with the feathered spears that had killed more men than he could remember.

Kowe's life spanned the entire modern history of the Waorani, a tribe that had once terrorized all of Amazonian Ecuador. As a child, he escaped raiding parties by burying himself in the mud and breathing through the hollow shafts of reeds. As a young hunter, he tracked peccary and capybara, shot hummingbirds out of the canopy, learned to distinguish by scent animal urine at forty paces and identify the species which had left it behind. With his father, he cleared fields using stone tools, ax heads not made by the people but found on the forest floor, the gifts of the creator Waengongi. He was an adult in 1957, when five missionaries died in an attempt to make first contact. Among their many mistakes, the American Baptists dropped from the air eight-by-ten, black-and-white glossy photographs of themselves in poses that we would view as benign and friendly: a warm smile, a friendly wave, a sincere look of concern. The Waorani, who had never seen anything two dimensional in their lives, lifted these pictures from the ground and looked behind the image, seeking the form of the face portrayed. Finding nothing, they concluded that these were messages from the devil, and when the missionaries arrived at a sandbar on the banks of the Río Curaray, promptly speared them to death.

After that, things went poorly for the Waorani. The military intervened, and the missionaries redoubled their efforts. In their isolation, the Waorani had been astonishingly healthy. Medical studies at the time of contact revealed a people essentially disease-free, with no history of cancer or heart ailments, and no evidence of exposure to polio, pneumonia, smallpox, chicken pox, typhus, typhoid fever, syphilis, tuberculosis, malaria, serum hepatitis, or the common cold. They had practically no internal parasites and virtually no secondary bacterial infections.

But in the wake of sustained contact, pestilential diseases swept the villages. Scores of Waorani succumbed to polio and influenza, afflictions deemed by the missionaries to be the judgement of God. When I met him, two decades later, Kowe was an elder lost between worlds. By night, he was a man of the forest, seeking inspiration in the sounds of the wild. By day, he dutifully attended the missionary church, mouthing the hymns, mimicking the motions of the younger members of the congregation. His own children and grandchildren had left to seek work with the oil companies that had laid claim to the vast pool of petroleum that by cruel chance lay beneath Waorani lands. From their labor came clothes, shotguns, flashlights, a cargo of goods that had undoubtedly made life easier.

Having been raised in complete isolation, Kowe, like all Waorani of his generation, had no sense of what it meant to lose his culture. By the time any of the Waorani understood what had transpired in but two decades of contact, the attraction of the new life was overpowering, and the only people who wanted to retain the old ways were the ones who had never lived it. Whatever else it had wrought, Christianity had stopped the spearing raids, the killing of innocent women, infanticide, and the live burial of children. Before contact, life as a Waorani had been many things, but pleasant was not one of them.

By tracing kinship through time, anthropologist Jim Yost discovered the extent to which warfare had dominated the lives of the Waorani: over the last five generations, no less than 54 percent of all the men, and 40 percent of the women, had died as the result of spearing raids. One in five had been shot or kidnapped by outsiders, the *cowade*, whom the Waorani believed to be cannibals. Astonishingly, over 5 percent of the effective mortality was due to individuals fleeing to the lands of the cowade, never to return. Presumably, they felt that life even among cannibals was preferable to the world they knew. Another 5 percent succumbed to poisonous snakes, the highest rate of such mortality recorded for a human population; 95 percent of Waorani men had been bitten by a venomous snake, 50 percent of them more than once. In his seven years with the tribe, Yost heard only of three instances of what he might consider natural death: the Waorani implied that the individuals in question had grown old and passed away. Then, one day, a young Wao inadvertently let slip that one of the men had "died becoming old"; that is, he had grown so old that the people decided to spear him and throw his body into the river.

Forty years after contact, the world that Kowe knew as a young man no longer exists. Spearing raids are a thing of the past and even the missionaries are gone, expelled by the Ecuadorian government. A people who but two generations ago employed stone tools to clear their fields are today participants in an ecotourism industry that each year brings scores of outsiders to their once isolated settlements. Waorani men work for oil companies, while their women barter in sunglasses, T-shirts, radios, baseball hats, and other trade goods. Healers conduct seminars in the forest, while tribal leaders attend conferences around the world, sponsored by multinational organizations. In the remote reaches of their territory, there may still be a small band of uncontacted Waorani, splintered from the majority, running scared in the forest. Whatever their fate, they clearly do not represent the future of the people.

In many ways, the plight of Waorani and their situation today exemplify the contradictions and challenges faced by modern anthropology. It is often said that anthropologists seek to deny indigenous peoples a chance to change, preferring instead to sequester them in static enclaves like some sort of zoological specimen. In truth, no serious anthropologist seeks to freeze a people in time. And no one who understands the life once led by the Waorani would wish it on anyone. Cultural survival is not about preservation. Change itself does not destroy a culture, since all societies are constantly evolving. Indeed, a culture survives, as David Maybury-Lewis has written, when it has enough confidence in its past and enough say in its future to maintain its spirit and essence through all the changes it will inevitably undergo. The Waorani did not stop being Waorani when their traditional cycle of self-destructive violence was broken. It was not the medical work of the missionaries, for example, that transformed Waorani life; it was the baggage that came with it, the imposition of the spiritual worldview of outsiders who believed that they had the monopoly on the route to God. Add to this the industrial exploitation of the Waorani homeland: the roads that pierced the wild, the pools of burning oil that in time would turn the emerald forests into a wasteland. The result is the bewilderment of Kowe, and the loss of another possibility of life.

~

IN 1990, MALAYSIA exported 9.4 billion cubic feet of tropical timber, over 90 percent in the form of raw logs, cut for the most part on Borneo in the forests of Sarawak, homeland of a dozen indigenous cultures, including the Penan, one of the last nomadic peoples of Southeast Asia. In the early 1980s, at a time when growing international concern focused the world's attention on the destruction of the Amazon rain forests, Brazil produced less than three percent of global tropical timber exports; Malaysia accounted for 60 percent. According to the World Bank, trees were being harvested in Sarawak at four times the sustainable rate. Industrial logging in the region has a shallow history, a mere four decades. But by 1993, in the Baram River

Kowe preparing poison darts, Quiwado,
Ecuadorian lowlands, 1981

Dawn in the forests of the Waorani,
Amazonian lowlands, 1981

drainage alone, there were more than thirty logging companies, some equipped with as many as twelve hundred bulldozers, working a million acres traditionally belonging to the Kayan, Kenyah, and Penan. By 1998, over 70 percent of Penan territory had been officially designated for commercial exploitation; much of it had already been logged.

As the forests fall, the indigenous cultures suffer the most, the very people who over the course of generations have developed an intimate knowledge of the land, men and women who, lacking the technology to transform the forest, chose instead long ago to understand it. In myth and stories, they celebrate the beauty of a landscape whose biological diversity astounds the scientific world. An entomologist working in Borneo identified some six hundred species of butterflies and caterpillars in a single day. Another reported over a thousand species of cicadas. In a collection of four thousand fungi, a mycologist found that half were unknown species. The floristic diversity surpasses that of the most prolific areas of the Amazon: in a series of forest plots encompassing in total a mere twenty-two acres, a botanist counted over seven hundred species of trees, as many as have been recorded for all of North America.

The ultimate fate of these forests remains unknown, but with the value of timber exports from Sarawak alone surpassing a billion dollars a year, the only certainty is that the logging will continue. I first traveled to Borneo and the Penan homeland in 1989. On my second visit in 1993, I found that logging roads had reached remote areas that had seemed impossibly distant from any threat but four years before. One afternoon, a Penan friend took me to a mountaintop where for generations his people had come to pray. From the summit, we looked out over a pristine forest, past the clear headwaters of one of Sarawak's ancient rivers to distant mountains that rose toward the heart of Borneo. There on the horizon, coming over the mountains and through the valleys from seven directions, were the raw scars of advancing logging roads. The nearest was but three miles from the encampment where we were staying. When the wind was blowing, we could hear the sound of chain saws, even as the light failed and night came upon us. Five years later, I returned to Sarawak again to assess the situation of the Penan.

THE BARAM RIVER is the color of the earth. To the north, the soils of Sarawak disappear into the South China Sea, and fleets of Japanese freighters hang on the horizon, awaiting the tides and a chance to fill their holds with raw logs ripped from the forests of Borneo. A hundred miles upriver, on the banks of the Tutoh at the Penan settlement of Long Iman, my old friend Mutang was away hunting wild pigs, to be sold to loggers in a nearby camp. His father, Tu'o, headman of the longhouse, was born in the forest, at a time when nearly all Penan were nomads, hunters and gatherers moving through the vast and remote forested uplands of northern Borneo.

All of Mutang's life has been marked by the frenzy of logging that has gripped Malaysia over the past four decades. Deprived of their traditional basis for life, the Penan drift toward government settlements built with the intent of drawing the people out of the forests. As a result, no more than three hundred of the roughly seven thousand Penan are today nomadic.

Long Iman is bleak, a wooden longhouse roofed in zinc, with great empty rooms and shuttered windows that keep out the wind. The river is soiled with silt and debris, the water no longer fit to drink. The afternoon rains turn the clearings around the settlement to a mud mire where children play. When I met Tu'o on the landing by the river, he greeted me warmly, touching my hand and passing his fingers over his heart. I didn't speak his name and he did not say mine. We called each other *padée*, brother, the proper salutation.

In the evening, children gathered around a television

and watched as a Malaysian journalist read the news in a language few of them understand. Tu'o apologized for the spartan fare at dinner: rice and broth, a plate of wild ferns. "How can you feed your guests in a settlement? It's not like the forest, where there is plenty of food. In the forest, I can give you as much as you want. Here, you just sit and stare at your guests, and you can't offer them anything. This longhouse is well built, and we have mattresses and pillows. But you can't eat a pillow."

Thirty years ago, government agents induced Tu'o to settle at Long Iman. Facilities that were promised—schools and clinics—were never built. There are few jobs, mostly menial work in logging camps, and with little experience, the Penan make poor farmers. For Tu'o, recalling the past is not a matter of mere nostalgia. It is a longing for a time when his children did not have to go to sleep hungry and his people lived by the grace of the forest, unaware of the impending cataclysm.

As we sat in a circle on the wooden slats of the kitchen floor, sharing food, I explained to Tu'o the purpose of our journey. Our goal was to reach one of the last bands of nomads, a cluster of families from the Ubong River who live for the most part in the remote reaches of Gunung Mulu National Park, a mountainous refuge that rises from the Tutoh River. Only in the confines of the park is the forest pristine and the traditional subsistence base intact. With me was my friend Ian Mackenzie, a Canadian linguist intent on compiling the first Penan grammar and dictionary, a labor of years. This was Ian's tenth sojourn among the Penan.

Asked to form an expedition overnight, Tu'o was ready by dawn. With six young Penan as our companions and Tu'o as guide, we left Long Iman in late morning and traveled up the Tutoh by longboat to reach a trail that climbed steeply through gingers and wild durian. Movement toward the ridge was slow and deliberate. For the Penan, even those recently settled, the destination is everywhere and nowhere. The capacity for survival lies all around in the forest.

Approaching the ridgeline, two hours above the Tutoh, we heard the low drone of logging trucks downshifting on the far side of the river. Peering through an opening in the trees across the narrow valley, Yapun, one of the young Penan, said with obvious contempt, "That is the work of Taib." Having denounced Abdul Taib Mahmud, the government minister responsible for Sarawak's forestry practices, Yapun turned his attention to a parade of insects at his feet. "If only we were as plentiful as these ants, the Malaysians would leave us alone. But it is they who outnumber the ants and make our lives so miserable."

For two long days, we walked into the forest, following a route that rose and fell with each successive ridge. Delighted to be away from the settlement, the Penan watched the forest for signs, hunting hornbills at dusk, tracking deer and sun bears, gathering the ripe fruits of feral mango trees. When, on the third morning, our party crested a steep hill and found a message tacked to a tree, we knew the nomads to be near. The note, scratched in broken Malay, said very simply that all the families had gathered to await us at an encampment known as Lamin Sapé.

~

IT WAS JUST AFTER DAWN and the sound of gibbons ran across the canopy of the forest. Smoke from cooking fires mingled with the mist. A hunting party returned. Tu'o bowed his head in morning prayer: "Thank you for the sun rising, for the trees and the forest of abundance, the trees that were not made by man, but by you."

On this ridge at Lamin Sapé, where generations of Penan have come, four families remained, in flimsy thatch shelters made of poles and rattan perched above the forest. Asik Nyelit is the headman. Ten years ago, I first met him on the banks of the Tutoh, when the river ran clear. He had recently been jailed for participating in the blockades that had shut down logging in much of Sarawak. Begun as a quixotic gesture, blowpipes confronting bulldozers, these protests electrified the international environmental movement, leading then Senator Al Gore to describe the Penan as the frontline troops in the battle to save the Earth. But the logging continued.

"From the time of our origins," Asik lamented, "we have preserved the trees and animals, every single thing in the forest. This we know. It is in our legends, our traditions. When we think of the places and our land, our hearts are troubled. Everywhere I go, I feel the need to weep."

For the Penan, the destruction of their forests represents far more than the loss of subsistence: it implies the death of a people. The forest is their homeland, and all their history is recorded in the landscape. Peter Brosius, an anthropologist at the University of Georgia, lived among the Penan for nearly four years. "The land," he says, "is filled with cultural significance. For streams alone, they have over two thousand names, each imbued with its own history. Logging does not merely destroy the trees and the forest, it destroys the things that are iconic of Penan society. Bulldozers and roads obliterate recognizable features. Once the canopy is opened, an impenetrable mass of thorny underbrush makes access and movement impossible. The cultural resonance of the landscape, all the sites with biographical, social, and historical significance, are hidden, producing a sort of collective amnesia."

In the morning, we went with the children to gather fruit in the forest. Pajak, the eldest of the nomads, who vows never to enter a settlement, sent with us two of his daughters, Tudé and Lesevet. The wild tangle of trails and vegetation would have lost us in a moment had we not been able to follow the nimble steps of girls not yet a decade old. They skipped across ravines, slipped laughing past thorn palms, squealed with delight as they climbed, hand over hand, up lianas that led to branches of white *langsat* fruits, sweet as nectar, which, shaken from the limbs, were gathered from the forest floor and carried back to camp in baskets woven from rattan.

Asik was there to meet us, and we followed him and his wife, Juna, down a steep and slippery slope to a glade in the forest where water runs. With him was his son Péndi, a toddler of two, scampering alongside a pet monkey, a pigtailed macaque that struck away at every opportunity. "Come back, my friend," Asik cried, "I will protect you." The monkey returned and leapt onto his shoulder. Though content to kill most anything in the wild, the Penan never harm an animal once it has been brought into the circle of the family. Nothing horrifies them more than the thought of raising domesticated animals for slaughter.

Along the trail, Asik pointed out leaves that heal, others that kill, and magical herbs believed to empower hunting dogs and dispel the forces of darkness. There are trees that produce rare resins and gums for trade with itinerant merchants, vines that yield twine and fiber for baskets, a liana that smolders for days and allows for the transport of fire.

The most important plant of all is the sago palm, the tree of life. Already that morning, Asik had cut down a stand. Now, on a bed of fresh leaves, he split the sections of trunk in half lengthwise and, with a slow steady rhythm, pounded the soft pith. Leached with water, the pith yields a thick paste, which is dried into sago flour, the main staple of the nomads. In an afternoon, Asik and Juna secured enough food for a week.

Two nights later, close to dusk, we sat by a fire as Tu'o cooked the head of a wild pig. Everything we hear, he explained, is an element of a language of the spirit. Thunder is the embodiment of *balei ja' au*, the most powerful magic in the woods. Trees bloom when they sense the song of the peacock pheasant. Birdcalls heard from a certain direction bear good tidings; the same sounds heard from a different direction are a harbinger of ill. Entire hunting parties can be turned back by the call of a banded kingfisher, the cry of a bat hawk.

Asik emerged from the forest, his face badly scratched and bleeding from an encounter with a thorny vine as he hunted monitor lizards in the canopy. Tu'o laughed as Asik recalled his folly: an entire day on the trail and nothing to show for it. Asik's nephew Gemuk appeared. Surprised to find us in his home, he poured a basket of rambutans at our feet, fruits that had taken hours to gather. Other Penan returned with baskets of *buaa nakan* fruits to roast, wild mushrooms for soup, hearts of palm, and succulent greens. Sharing for the Penan is an ingrained reflex. When Gemuk announced that not one but two wild pigs had been killed, Asik roared with delight. "Don't be hungry. Good to be full."

Moments later, there was a shout from the ridge. Two other families had arrived, just as a tremendous thunderstorm cracked open the sky. In the midst of the downpour, they erected *lamin*, shelters built in an hour that will house them for a month. The men and boys cut poles and rattan to build the frame; the women gathered palm leaves to be sewn into thatch. A fire was kindled. Infants huddled beneath leaves while older children assisted their parents. Lakei Padeng is Asik's stepfather, an old man known as "black face." I watched as he emptied the rattan backpacks. Two families—five adults, eleven children—together possessed a kettle, a wok, several sharpening stones, dart quivers and blowpipes, sleeping mats, an ax, a few ragged clothes, a tin box and key, two flashlights, a cassette player, three tapes, eight dogs, two monkeys.

"The Penan are so profoundly different," Ian remarked later that night. "They have no writing, so their total vocabulary at any one time is the knowledge of the best storyteller. There is one word for 'he,' 'she,' and 'it,' but six for 'we.' There are at least eight words for sago, because it is the plant that allows them to survive. Sharing is an obligation, so there is no word for 'thank you.' They can name hundreds of trees but there is no word for 'forest.' Their universe is divided between *tana' lihep, tana' lalun*—'land of shade,' 'land of abundance'—and *tana' tasa'*, 'land that has been destroyed.' "

Language provides clues to a complex social world utterly different from that of sedentary peoples. The nomadic Penan have no sense of time, know nothing of paid employment, of poverty. They have no notion of work as a burden, as opposed to leisure as recreation. For them, there is only life, the daily round. Children learn not in school but through experience, often at the side of their parents. With families and individuals dispersed much of the time throughout the forest, everyone must be self-sufficient, capable of doing every task. Thus, there are no specialists and little hierarchy. As in many hunting societies, direct criticism of another is frowned upon, for the priority always is the solidarity of the group. Should conflict lead to a schism and families go their separate ways for prolonged periods, both groups may starve for want of sufficient hunters.

The greatest contrast between the Penan and ourselves may well be the value that they place on community. Since they carry everything on their backs, they have no incentive to accumulate material objects. They measure wealth not by the extent of their possessions but by the strength of their relationships. It is the simple result of adaptation, though the consequences are profound. In our tradition, we long ago liberated the individual, a decisive shift in orientation that David Maybury-Lewis has described as the sociological equivalent of splitting the atom, for in doing so, we severed the obligations of kin and community that, for better and for worse, constrain the individual in traditional societies. In glorifying the self, we did away with community. The consequences we encounter everyday in the streets of our cities. An American child grows up believing, for example, that homelessness is a regrettable but inevitable feature of life. A child of the nomadic Penan, by contrast, is taught that a poor man shames us all.

"In the place they want us to live," Asik said, referring to the Malaysian policy of encouraging the Penan to settle in places like Long Iman, "the sago is gone, the trees have been destroyed, and all the land is ruined. The animals are gone, the rivers are muddy. Here we sleep on hard logs, but we have plenty to eat."

Addressing a meeting of European and Asian leaders in 1990, Prime Minister Mahathir Mohamad of Malaysia remarked: "It is our policy to eventually bring all jungle dwellers into the mainstream ... There is nothing romantic about these helpless, half-starved, and disease-ridden people."

"We don't want them running around like animals," said James Wong, Sarawak's minister for housing and public health. "No one has the ethical right to deprive the Penan of the right to assimilation into Malaysian society."

It is a struggle to reconcile such a statement with the desolate reality of Long Iman, the image of Asik in the forest, or the memory of his children frolicking in the clear stream beneath the shade of giant ferns. While a number of individual Penan have benefited from the economic development of the past thirty years and the population has more than doubled, largely because of improvements in basic health care, for the majority in the longhouses, there has only been impoverishment. Throughout the homeland of the Penan, the sago and rattan, the palms, lianas, and fruit trees lie crushed on the forest floor. The hornbill has fled with the pheasants, and as the trees fall, a unique vision of life is fading in a single generation.

WHEN I RETURNED from Borneo, I spoke with David Maybury-Lewis, who, like so many anthropologists, was both appalled by the policies of the Malaysian government and deeply sympathetic with the plight of the Penan, which he viewed as symptomatic of a global dilemma. "Genocide, the physical extermination of a people, is universally condemned," he noted, "but ethnocide, the destruction of a people's way of life, is not only not condemned when it comes to indigenous peoples, it is advocated as appropriate policy."

The Malaysians want to emancipate the Penan from their backwardness, which means freeing them from who they actually are. Indigenous peoples such as the Penan are said to stand in the way of development, which becomes grounds for dispossessing them and destroying their way of life. Their disappearance is then described as inevitable, as such archaic folk cannot be expected to survive in the modern world.

The idea that indigenous societies are incapable of change and bound to fade away is wrong, according to Maybury-Lewis. What needs to be considered is the very notion that nations have an inherent right to do what they choose to ancient peoples within their boundaries. Malaysia, like many modern countries, was formed from the residue of a colonial empire, and in the days before the British, the land now called Sarawak, located five hundred miles across the South China Sea, had few ties with the states of the Malay Peninsula. The Federation of Malaysia is not yet forty years old, and this time span does not, in any moral or ethical sense, grant it the right to abuse, in the name of national sovereignty, human rights and natural resources of global significance.

Malaysia is but one example. Indonesia, a nation of fourteen thousand islands with a population of more than 200 million, has three hundred ethnic groups whose only common historical experience was Dutch rule. For most inhabitants of the vast archipelago, independence meant only that one colonial master was replaced with another. Those living in the outer reaches of Indonesia, in Irian Jaya and Sumatra, Timor and the Moluccas, want little to do with Javanese rule, and efforts to establish a national presence through forced migration and other government programs have sparked ethnic violence throughout the land, most notably in Kalimantan in Indonesian Borneo.

The limestone pinnacles of Gunung Mulu,
Sarawak, 1989

Dawn at Long Sabai, a Penan settlement on the upper Tutoh River, Sarawak, 1993

Tu'o Pejuman, Asik Nyelit, and Tingan Déng at Lamin Sapé, Sarawak, 1998

Facing page: A Penan boy brings home the head of a bearded pig, Lamin Sapé, Sarawak, 1998

The Penan settlement of Long Iman, Sarawak, 1993

A logging camp on the Baram River, Sarawak, 1993

The entire map of Africa is delineated by boundaries, many of which have no historical resonance and reflect only the arbitrary legacy of colonialism. Wars rage over much of the continent as power shifts between local factions, military thugs for the most part, who manage to momentarily control the local flow of wealth, gold, diamonds, oil, and thus secure access to arms. National governments claim to speak for peoples they have no moral or political justification to represent. International organizations, keen to offer aid, yet often motivated by arbitrary agendas, apply solutions to situations they little comprehend, invoking remedies for problems not infrequently of their own making.

"In so many ways Daniel Bell was right," Maybury-Lewis added, referring to the well known Harvard sociologist: "The nation state has become too small for the big problems of the world, and too big for the little problems of the world. Too often we meddle with lives we barely understand."

Like so much of what he had told me over the years, I carried these words away, certain that one day they would come back to me, as indeed they did, the first time I traveled to East Africa.

STUDDED WITH CRATER LAKES and blanketed by lush forests that are home to the largest elephants in Africa, Mount Marsabit stands as a fertile sentinel above the barren sands of northern Kenya. To the east, a flat horizon reaches to Somalia. To the south and west lies the Kaisut Desert, and beyond are the Ndoto Mountains, a rim of peaks rising more than 8,500 feet above the white heat of the lowlands. For thousands of years, pastoral nomads thrived here because they and their animals traveled lightly on the land. Mobility was the key to survival. Drought, the long hunger that descends ruthlessly from a searing sky, was not a cruel anomaly but a constant if unpredictable feature of life and climate. Surviving drought was the essential challenge that made the desert peoples of Kenya who they are: Turkana and Boran, Rendille, Samburu, Ariaal, and Gabra.

In the wake of a series of devastating droughts in the 1970s and 1980s, along with famine induced by ethnic conflict and war in neighboring Ethiopia and Somalia, international organizations arrived by the score to distribute relief. Mission posts with clinics, churches, schools, and free food drew the people from the parched land. At the same time, and despite evidence to the contrary, it became accepted in development circles that the nomads themselves were to blame for degrading their environment through overgrazing. In 1976, the United Nations launched a multimillion-dollar effort to encourage two of the tribes, in particular, the closely related Rendille and Ariaal, to settle and enter a cash economy, reducing the size of their herds by selling stock. This dovetailed with the interests of those Kenyans who considered nomads a symbol of the past and saw education and modernization as the key to the country's future.

For the ten thousand Ariaal herders, circumstances for the most part were not so dire that they were forced to settle. Neither fully Samburu nor Rendille, the Ariaal are a remarkable fusion that emerged in the late nineteenth century. Decimated by famine, a splinter group of Rendille, camel herders from the Kaisut Desert, moved to the marginal lands in the shadow of the Ndoto Mountains, where they established close relations with the Samburu, a cattle-raising people of the highlands. Adopting the ways of the Samburu yet retaining many Rendille customs, speaking both languages, the Ariaal had the best of both worlds. On the western flank of Mount Marsabit and in the Ndoto foothills, where water could almost always be found, they kept cattle, while far below in the desert, their camels foraged in the shade of frail acacias. As Kenyans say, the Ariaal have the bones of Rendille, but their meat is Samburu. Thus, they secured their survival.

The Rendille, by contrast, thirty thousand strong and totally dependent on their camels, suffered terrible losses in the droughts and drifted by the thousand toward the relief camps. By 1985, more than 75 percent of the tribe lived in destitution around the lowland towns of Korr and Kargi, their well-being inextricably linked to mission handouts.

~

WRAPPED IN A RED SHAWL, an old man with a wizened face and earlobes studded in gold reached for my hand and nonchalantly spat into the upturned palm. "It's a sign of greeting," explained Kevin Smith, a young American anthropologist, as we walked with one of his clan brothers, his closest Ariaal friend, Jonathan Lengalen, through Karare, a community on the southern slope of Mount Marsabit. With a full moon over the grassland, Jonathan led us along a chalky trail to his *manyatta*, a cluster of domed shelters built of branches and mud, cow dung and hides. From the shadows emerged the warriors, tall and thin, their long hair woven in tight braids dyed red with ocher and fat. Their bodies shone with decoration. All carried weapons, swords sheathed in leather, wooden clubs, iron spears, the odd assault rifle. Already, they were singing, deep resonant chants that drew the young girls, equally beautiful in beads and ocher, into the clearing. As the warriors moved forward, slapping the girls with their hair and leaping into the air with the grace of gazelles, their spears flashed in the moonlight.

The singing and dancing lasted well into the night. With the end of the rains, grass was abundant and milk plentiful. It was a time of great joy, a season of celebrations, and almost every day there was a wedding. Sunrise found us in a cool mist, walking with Sekwa Lesuyai and his best man as they led a bull and eight heifers along a trail that climbed toward the home of his bride, Nantalian Lenure. All night, the men had slept beside the animals, guarding the gifts that would secure the marriage. The bride's mother washed the men's feet with milk. The bull was slaughtered, its meat distributed with ritual precision. The elders brewed tea and then slipped away from the manyatta into the bush to roast and eat their share of the meat. The women stayed by Nantalian and laid branches of fertility across her doorway. "God is big," they sang, "big as a mountain; the bride is beautiful, sweet as perfume." Only in the late afternoon did the warriors arrive, to resume their dancing with an intensity that drove several of them into trance.

Two days later, Jonathan invited us to spend a night in one of the remote encampments, or *fora*, where the warriors live apart with the young lads, managing and protecting the herds, raiding enemy tribes. For ten years, from the time they are circumcised until they are finally permitted by the elders to marry, they spend each night in the open, sleeping on the stony ground, living on soups made of wild herbs, and on fresh milk and blood, drawn each evening from the neck of a heifer. Sitting with the warriors in the moonlight, the sound of cattle bells ringing in the night and the friendly faces of Zebu cows crowding the fire, I came to understand how for the Ariaal these animals are the fulcrum of life.

When men meet on a trail, they ask first of the well-being of the herds, then of the families. Each animal has a mark and a name, a personality setting it apart. Cattle represent a man's wealth and status, and without herds, he cannot marry. But the bond is deeper, even spiritual, rooted in every intuition about the landscape and environment.

"If we lose our cows," said Jonathan," we lose our faith in life itself. All our rituals and ceremonies lose their meaning without the animals."

He reached for a burning stick, snapped off the ember, and dropped it into his empty milk gourd. In the absence of water, it is the way the Ariaal clean their containers. As Jonathan spoke of his tribe, the outline of the culture unfolded in my mind.

Everything is built around the need to manage risk. The larger the herd, the greater the chances that some animals will survive a difficult period. Keeping the herd intact is essential. Thus, with the death of an owner, the eldest son inherits everything.

The need to care for hundreds of animals creates an incentive to have many children and wives to help with the work. Polygyny addresses this problem but inevitably creates tensions within the society. With the old men having three and sometimes four wives, there is a shortage of marriageable women for the young men. This dilemma the elders solve by getting rid of the young men, sending them off to warrior encampments. But to make their exile desirable, it is enveloped with prestige.

The highlight of a young man's life is his public circumcision, the moment when he and his peers enter the privileged world of the warrior. The ceremony is held only every fourteen years, and those who endure it together are bonded for life.

"You sit perfectly still," Jonathan remembered, "legs apart, with your back supported by your closest friend. They pour milk on you. Everyone is singing or yelling, warning you not to flinch. All your family promises animals, if you are brave. You can build up a herd just with those frantic promises. But you are so intent. You only hope that the blade is sharp. It's over in seconds, but it seems like years." Should a boy, his head shaved and blackened with fat and charcoal, reveal the slightest expression of fear or pain as the nine cuts are made to his foreskin, he will shame his clan forever and possibly be beaten to death. Few fail, for the honor is immense.

"God has given us our land," Jonathan said, "the land that we share. Our traditions we have created and they are our strength. As long as we have land and cattle, and respect for the elders and the past, we will have our culture."

After ten days on Mount Marsabit, Kevin Smith and I drove to the desert lowlands to visit Lewogoso and Losidan, Ariaal nomadic encampments along the base of the Ndoto Mountains, and then on to the town of Korr, the mission post where so many Rendille have settled. The contrast between the two worlds could not have been greater.

In the Ariaal manyattas, traditions were strong and enduring, embraced consciously by the people. On rocky outcrops, warriors painted in ocher stood like raptors overlooking the narrow traffic of camels and cattle on desert trails. Elders gathered each night in the *na'abo*, the ritual men's circle, to offer prayers. At Lewogoso, women and children were herding goats and sheep, and drawing blood in the morning from the faces of camels. At Losidan, a death had occurred. The manyatta was deserted, though the cooking fires were still warm. By custom, the people had moved on. Following them into the desert, we met Kanikis Leaduma, a *laibon*, a healer and soothsayer, who reads the future in colored stones and bones tossed from a gourd onto a green cloth spread out in the shade of an acacia tree. A young man of perhaps twenty-five, he had acquired the gift of clairvoyance from his father and discovered in dreams the secrets of health and well-being. Fighting sorcery with amulets and herbs, he protected livestock and people while providing an anchor of spiritual certainty in a harsh and unforgiving desert.

"If they can control their land and maintain their pastoral economy, the culture will thrive," Kevin remarked as we drove along the faint outlines of a desert track that led from Lewogoso to Korr. To understand the plight of the desert tribes, he suggested, one had to begin with the nation state, and the convictions and biases of sedentary people for whom nomads are an inconvenience. Highly mobile, straddling international borders, living on the margins of the world, nomads are envied for their freedom and independence, and hated and feared for these same traits. They pay no taxes, are beholden to no government, move at will across landscapes that mock the arbitrary borders scratched upon the face of post-colonial Africa.

In the 1970s, drought and famine drew the attention of the world to the sub-Sahara. The development community insisted that the degradation of the Sahel and the impoverishment of the people were the inevitable consequence of a pathology academically described as the "tragedy of the commons." As long as nomads were free to exploit the desert at will, the argument suggested, individual greed and the desire to maximize personal economic gain would inevitably triumph over the interests of the community, resulting in overgrazing and the erosion of the land. The solution was privatization and the imposition of a model of land tenure, fenced rangelands and all, imported wholesale from the ranches of the American West. The audacity of such an alien prescription was nothing new. Since the arrival of the British in East Africa, the nomadic peoples have been told how to manage their lands by outsiders, missionaries, government officials, foreigners of every color.

Yet, for thousands of years, the very survival of the nomads had, by definition, depended upon their looking after the land. The desert is their home, a place of freedom and fertility, of good grasses and bad, of protective trees and hidden springs. Using animals to convert scrub vegetation to protein is not only the most efficient use of the land, it is the only way to live in the desert. Mediating the process, securing the rights of every individual through linkage to the fate of the collective, were complex ties of kinship, relationships too subtle to be readily perceived by outsiders, especially those blinded by hubris. Through generations, the nomads had discovered the art of survival in the desert.

Grazing and the deposit of animal waste returns nitrogen to the ground, enhancing the growth of grass. Lands overgrazed for a short period produce richer fodder in the wake of the herds, as gravel and seeds are crushed by hooves. Different springs have different kinds of water. The mineral content varies. The nomads recognize this and seek the appropriate water for the time of year, releasing nutrients from the deep wells to the surface of the land. Medical surveys reveal that the milk and blood diet of the nomads is far superior to the food available in the missionary towns, and that their children are healthier, despite the lack of Western health care. The problems begin, for the most part, when a people born to move settle down.

As recently as 1975, Korr was a seasonal watering hole visited by small bands of nomadic Rendille herders. That year, Italian missionaries set up a small camp to distribute relief. Within a decade, a town grew, complete with shops, schools, and a large stone Roman Catholic church. Today, there are twenty-five hundred houses in walking distance of the mission, a local population of sixteen thousand, and 170 hand-dug wells. Missing are the trees that once provided shelter in a windswept desert. Most have been cut down to produce charcoal to cook maize gruel, the staple subsidy. Those Rendille who still own camels and goats must herd them far from town. Fresh milk is hard to find, and many children go without. In place of sisal, the houses are roofed in cardboard, burlap, and metal sheets bearing the names of international relief agencies. A walk around town reveals that almost every Western nation has helped create this oasis of dependency.

For many Rendille, it was not just food but a chance to educate their children that drew them to Korr. In the face of drought, having at least one child in school destined for the cash economy was another means of managing risk. One who made that choice was our host, Kawab Bulyar Lago. Born in the desert but crippled by polio in his teens, Kawab grew up in a mission and became one of the first of his tribe to be educated. Kawab, who had seven children, sold his animals and saddled himself with debt to send his eldest son, Paul, to Catholic school outside Marsabit. Awaiting the results of the national exams that will determine his fate, Paul hopes to attend university and become a doctor, a civil engineer, a teacher, or even a tour guide. "I'd prefer to be a doctor," he told me one morning, "but anything would be all right."

As I listened to his stories of school, with its curriculum dominated by Western religious studies, of the dormitory teasing he endured that led his father to have him circumcised in a hospital before his time, I sensed his anxiety and could not help but recall the calm authority and confidence of Kanikis the healer, a young Ariaal man of similar age so firmly rooted in tradition. In the old days, Paul would have inherited his father's herd. Today, he inherits his hopes and dreams. The family has staked everything on his education. Kawab knows the risks. "For the few who benefit, who get good jobs, they do well," he said. "But most just suffer. Here, you can survive. People will help you. In the city, they just leave you to die."

Education, long held out as the key to modernization, has its critics. Father George, the Catholic priest who has run the mission in Korr almost since its inception, told me: "Schooling has not changed people for the better. This is the pain in my heart. Those educated want nothing to do with their animals. They just want to leave. Education should not be a reason to go away. It's an obligation to come back."

As we sat together in his modest kitchen, I perceived the old priest's sadness and disappointment. For twenty years or more, he had given everything to this windswept place scratched out of the desert. Hundreds of Rendille children had moved through his school. Many had drifted to the cities and larger settlements to the south. For them, education had been, by definition, a reason to go away, a passport out. Unfortunately, in a nation like Kenya, the destination is too often the bottom rung of an economic ladder that goes nowhere. Unemployment rates in Kenya's cities hover at around 25 percent. Among those who have attended school in the Marsabit District, well over half are unemployed.

The danger of Western-style education is clear to many Ariaal, including Lenguye, an elderly midwife from Ngurunit, a settlement in the foothills of the Ndoto Mountains. "We send our children to school, and they forget everything. It's the worst thing that ever happened to our people. They only know how to say 'give me.' They don't know how to say 'I give to you.' "

According to Nigel Pavitt, author of several books on the nomadic peoples of Kenya, "Education must be tailored to the needs of the people. They ought to be taught veterinary medicine, subjects useful to a livestock economy. At present the curriculum is designed to produce clerks. The methods are terrible. Boarding schools for nomads. They should be teaching teachers to be nomads."

"They must hold onto tradition," explained Father George as he guided me through the dark streets of Korr. "Ultimately, it is what will save them. It's all they have. They are Rendille and must stay Rendille."

In the end, the cultures that survive will be those that are willing and able to embrace the new on their own terms while rejecting anything that implies the total violation of their way of life. Seduced by empty promises, the Rendille took a chance and settled, a gamble that largely failed, leaving the bulk of the tribe wasted and abandoned. But at the fringe of the desert the Ariaal, despite droughts and famine and countless other pressures from within and without, have, for the moment, found a way to stay.

~

When we returned from the lowlands to Mount Marsabit, word reached us of a killing, a Rendille elder shot dead by a raiding party of Boran warriors as he tended his goats. The incident was described in a casual manner, just another flash of violence sparked by the whirlwind of change that has swept northern Kenya in the last generation. At one time the wealthiest pastoralists in the country, rarely in conflict with the Rendille, the Boran were brutally repressed in the early 1960s when they allied themselves with the losing side of a civil war. Their herds slaughtered, the men scratched a living making charcoal or gathering firewood and the women were forced to turn to prostitution. By 1971, a once proud people were starving and totally dependent on outside relief for survival. A decade later, their numbers began to grow, as other Boran in neighboring Ethiopia in turn suffered the ravages of famine and civil strife, and drifted south as refugees. Arriving on the northern flank of Marsabit, the Boran clashed in time with the settled Rendille, most notably those from Songa, an agricultural community established only in 1972 by an American missionary.

The first fatal skirmish occurred in December 1987, as Boran warriors came to Songa to water their cattle. The four dead, three Boran and one Rendille, were just a beginning. Within months, the blood feud had taken the lives of infants and women, old men and children. Retaliation became random, with the innocent dying on both sides. Within a year, Rendille and Boran were killing each other on sight. Automatic weapons increased the carnage. A Russian AK-47 could be had for the price of four bulls, a German G-3 for less. By December 1998, when I visited the region, it was no longer safe to travel from Songa to the market center at Marsabit. Three times a week, armed convoys set out on foot, a dozen escorts guarding three hundred or more women and donkeys carrying mangos and kale, papayas, oranges, and avocados in a wary procession that moved slowly along the dirt track that ran through the dense forest.

One day, I walked with them, along with my friend Daniel Lemoille, whose father had been among the first Rendille to settle at Songa. We set out at dawn, with low clouds clinging to the mountain, the women all around us brilliant in their scarves and blouses of bright reds and purples and yellows. Many in the convoy had been victims of the troubles. Nachapungai Lekupes's husband had been shot dead in 1996, leaving her with five children and pregnant. One of the armed escort turned out to be Sekwa Lesuyai, whose wedding I had attended but a week before. Dressed in black army boots and camouflage mufti, sunglasses in place of ocher, a rifle on his shoulder instead of a spear, he was hardly recognizable. Sekwa had lost a sister to the violence, one of seven killed in a massacre at Kukituruni in 1992. An elder named Paul Masokonte showed me the scar on his shoulder where he had been wounded; his friend, shot in the chest, had not survived the ambush. Only six days before, while I had been at Lewogoso watching camels being bled, an ambush had left three wounded: a policeman shot in the shoulder, an old woman in the thigh, and a young boy who lost two fingers to a bullet. "We will kill ten of them, for every one of us who dies," Daniel assured me.

Fortunately, it was a day without incident, and, as we approached the outskirts of Marsabit, the irrepressible spirit of the people burst forth in laughter and song, and a madcap rush to the town center as each woman secured her place in the market. In an instant, all tensions dissipated, and a cloak of normalcy enveloped the moment. War, famine, killing, the suffering and humiliations resulting from the ill-conceived schemes of foreigners: these things were forgotten, if only for a time. As I followed the women into Marsabit town, I had a profound sense that if only they had been left alone, if the opposing sides in the Cold War had not flooded a continent with guns, if all the missionaries and aid workers had kept their good intentions to themselves, then these ancient pastoral peoples might have had the opportunity to change at their own pace and in their own ways. Conflicts there may have been, but almost certainly they would have been less bloody.

Still, despite the violence and the lingering impact of development initiatives, the future of the nomadic peoples of northern Kenya remains far brighter than that of other peoples who have been overwhelmed by outsiders. The Chinese entered Tibet with the promise of liberation and the purported goal of transforming the lives of the Tibetan peasants for the better. What transpired, in fact, was a nightmare, as dark as any recorded in the bitter annals of human endeavor.

An Ariaal woman of Karare returning home with firewood, Mount Marsabit, Kenya, 1998

An Ariaal warrior and his cattle,
Mount Marsabit, Kenya, 1998

An Ariaal maiden at a wedding, Karare, Kenya, 1998

A mother and child at Lewogoso, Ndoto Mountains, Kenya, 1998

On the Songa Road, a Rendille armed escort
guards women on their way to market,
Mount Marsabit, Kenya, 1998

Facing page: An Ariaal youth tends
the fire in the warriors' encampment,
Mount Marsabit, Kenya, 1998

A warrior waters his cattle at the settlement of
Lewogoso, Ndoto Mountains, Kenya, 1998

An Ariaal elder visits a warriors' encampment,
Mount Marsabit, Kenya, 1998

A TEMPLE GUARDIAN AT TASHILHUNPO MONASTERY,
SHIGATSE, TIBET, 1997

CHAPTER SIX

THE LAND OF SNOWS

LATE ONE AFTERNOON IN LHASA, at the end of a long winter journey, a Tibetan friend led me through a warren of whitewashed houses, along a narrow frozen track that led eventually to the door of his mother's house. Content to be back in the ancient city, he sang as he walked, scattering coins to young boys wrapped in wool, faces black with soot, who followed in his wake, calling out his name. Upon reaching his home, we stepped beneath a pleated awning of red, blue, and yellow cloth, and entered a dark room scented with incense and illuminated by the faint glow of candles and butter lamps. Along two walls ran low benches cushioned with thick rugs. Painted chests served as tables. An iron brazier, glowing with burning juniper and yak dung, drew us to the middle of the room where we stood, hands to the flames, stooped beneath the low roof beams, ornately carved and stained a fiery red. In one corner of the room, beside an antique armoire, was a butter urn, its wood darkened and brass burnished from years of use. Above the urn hung a portrait of Mao Zedong and a photograph of the Potola Palace, the architectural wonder that was home to the Dalai Lamas and the seat of Tibetan political power until the Chinese invasion of 1950 and the crushing events that unfolded over the ensuing years.

My friend's niece squealed with delight to see him. She was a lovely girl, perhaps twenty, with walnut skin and black, finely braided hair that shone with oil. His mother was even more beautiful, a diminutive woman, hair streaked with gray, whose deeply lined face glowed with timid delight as she politely touched my hand and then embraced her son. Around her neck she wore necklaces of silver coins and coral, amber and turquoise, all that remained of the wealth that had once positioned the family at the highest level of Tibetan society.

As we drank butter tea, telling stories of our journey, laughing as we recalled moments of folly, my friend's mother sat close by his side, her weathered hands resting in his. In her long life, she had seen many things, endured unspeakable hardships. Her husband had been a confidant of the Pachen Lama, Tibet's second-most prominent religious leader, and thus had been murdered by the

Chinese following the revolt of 1959. Her brother, a high lama, had fled to India with His Holiness, the Fourteenth Dalai Lama, during the tragic diaspora. And she herself, condemned by association as a counterrevolutionary, a "stubborn running dog of imperialism" as the indictment had read, had been taken from her children and jailed. As an infant, my friend had been smuggled by an older sister into the prison, where for many months he lived hidden in the shadows beneath his mother's skirt. The brave sister was later sent to a re-education camp, where one day she inadvertently stepped on an armband, painted with characters extolling the virtues of Mao, that had slipped off the sleeve of an adjacent worker. For this transgression, she was condemned to seven years of hard labor.

Today, things are said to be somewhat better in Tibet, though the darkness of those years, a time when the sky fell to the earth, still haunts the land. When I mentioned that I had met the Dalai Lama and heard him speak, as have so many of my compatriots, my friend's mother began to cry. She reached to touch my hand, drawing my face close to hers, until her tears ran down my cheeks and I could taste the salt in the corners of my mouth. With a glance toward her children, she beckoned me to follow her into a small room, where a wall, responding to her pressure, opened to reveal a hidden devotional chamber. On the altar were brass urns, offerings of fruit and money, small piles of yak butter, and sticks of incense. There was a figure of the Dharma Chakra, the Wheel of Religion, supported by sacred deer represented in copper and gold, the symbol of the Buddhist path, life constantly in motion.

At the center of the shrine, a ceremonial scarf of blessing, a white silk *khata*, was draped around another photograph of the Potola. In this picture, there is a rainbow above the palace, and within the rainbow floats the face of the Dalai Lama, whose image is today banned by law throughout Tibet. At the risk of her freedom, this small and resolute woman continues to contemplate the Buddha mind and the great lessons of her faith. In better times, Milerepa, Tibet's beloved mystic saint, advised his students to "regard as one this life and the next and the one in between, and become accustomed to them all." Everything changes. Tibet was free, and then overwhelmed by the Chinese. The weak became strong. The ancient became youthful. Peace gave way to war, violence yielded to submission. This, too, will change. It is only a question of time.

In a bloodstained century, Mao Zedong bears the dark distinction of being the political leader most successful in killing his own people. In 1959, the year his army crushed the Lhasa uprising, Mao set in motion the Great Leap Forward. In order to decentralize and increase the production of steel, every farmer in China was ordered to construct a backyard furnace and obliged to meet imposed quotas. To do so, they melted their ploughs, reducing farm implements to ingots. For a year, China was statistically the top producer in the world. But then came the spring, and the fields lay fallow. In the fall, there was nothing to harvest. By winter, hunger stalked the land, and a famine unprecedented in this century took the lives of more than 40 million Chinese. Such was the wisdom and logic that marched into Tibet, intent on the destruction of a civilization.

Though often portrayed as an isolated land, a cosmic meritocracy sitting astride the heights of Asia, Tibet in truth was always a crossroads of trade and empire, a nation of great contradictions and inequalities. In the beginning, it was Tibet that had threatened China. For two hundred years, from the seventh century, Tibet's armies dominated Central Asia, controlling much of the Silk Road and absorbing into her empire large swaths of Chinese territory. In the ninth century, as political power in Tibet fragmented, the Buddhist religion, carried into the mountains from the lowland plains of India, emerged from the chaos as the pre-eminent esoteric vehicle for transformation and empowerment. As a spiritual force, Tibetan Buddhism moved far beyond the frontiers of the old empire, and wandering lamas, wizards in the realms of the spirit, found a place in the courts of kings. Within forty years of Genghis Khan's conquest of Tibet in 1207, Buddhist monks were providing religious instruction to the invaders.

China's modern claim to the land of snows, the coveted "treasure house of the west," is rooted in a questionable interpretation of this early history. In 1279, the Mongols, already in possession of Tibet, advanced under Kubla Khan into China, expanding their empire and founding the Yuan dynasty. The Chinese today trace their territorial rights to this moment when both nations succumbed to the Mongols. The Tibetans, by contrast, remember that distant era as a time when two sovereign nations fell in succession to a single enemy. When the Yuan dynasty collapsed, Tibet reclaimed its independence. Though political and religious ties remained, their strength ebbed and flowed. At no point during the subsequent years did Tibet become an integral part of Han China.

The Tibet that emerged over the centuries was not a perfect society, but its failures were its own. Looking south and east, fending off invasions, struggling with civil strife and intrigue from within, Tibetans engaged in the sordid realities of nationhood. When the Nepalese invaded in 1788, reaching Shigatse and looting the sacred monastery of Tashilhunpo, Tibet forged an alliance with China. The Qing emperor dispatched an army that vanquished the Nepalese and then, in the wake of victory, refused to leave. The Chinese maintained a modest presence in Lhasa until Qing influence faded in the late nineteenth century. By then, China faced its own enemies, European powers descending on Asia from the eastern sea. In the early years of the twentieth century, Chinese authority in Tibet was symbolic, marginal at best. Whatever claim they had ended in 1911, with the overthrow of the Qing dynasty. From 1913 until the invasion of the People's Liberation Army in 1949, Tibet was again an independent nation, albeit a complex land on the cusp of change. The very idea of China today claiming Tibet, a land and people unique by any ethnographic or historical definition, is as anachronistic as England laying claim to the United States simply because it was once a British colony.

Within a year of the establishment of the People's Republic of China, Mao's armies had swept through the eastern Tibetan provinces of Amdo and Kham, taking control of all regions outside the direct and historic jurisdiction of the Dalai Lama. Communist cadres sparked class struggle and instituted thought reform. In the name of liberation, communities were shattered, traditional ways forcibly abandoned, and the economy transformed through the collectivization of land and the seizure and redistribution of private property. Those who tried to resist, who tried to defend their families, fields, and temples, were imprisoned, tortured, and in many instances killed.

By the fall of 1950, thirty thousand battle-hardened Chinese troops—soldiers who had beaten the Japanese, defeated the Nationalists, and ridden a wave of victories that had placed all of China in Mao's hands—stood poised on the frontiers of inner Tibet. In a gesture emblematic of the contrast between the two worlds about to clash, the Regent and religious authorities in Lhasa turned to the State Oracles to ascertain the proper course of action. Asked whether resistance should be violent or non-violent, the Gadung oracle equivocated. Only by offering prayers and propitiating the deities, read the pronouncement, could the Dharma be protected. In the circumstances, the answer was clearly inadequate. Within days, Tibet's modest forces, equipped with limited arms and obliged to defend the entire frontier, were shredded by the Chinese.

When, in the spring of 1951, the United Nations, preoccupied with the conflict in Korea, rejected Tibet's appeal for intervention, the Dalai Lama had little choice but to resign himself to the 17 Point Agreement, a negotiated settlement signed in Beijing, calling for the "peaceful liberation" of his country. On the face of it, the document was moderate. In exchange for the return of all Tibetan people to the family of the motherland, the Chinese promised to respect religious and economic traditions and to grant complete regional autonomy to the heartland of Tibet. Hoping to spare at least some of his people the fate that had befallen the eastern regions of Amdo and Kham, the Dalai Lama acquiesced. Even as troops of the People's Liberation Army marched into Lhasa, doubling the population and precipitating massive shortages of food and fuel, he still sought an accommodation with the Chinese. In 1954, he traveled to Beijing to meet Mao and expressed a willingness to become a party member in order to work out, as he wrote, "a synthesis of Buddhist and Marxist doctrines." Such hopes of reconciliation proved illusory. To the communists, Tibet was a land of serfdom and exploitation, theocratic parasites and decadent aristocrats. The sooner they were gone, the better. The ultimate goal was to rid the nation of its backwardness. At his last meeting with the Dalai Lama, Mao dismissed a thousand years of Tibetan history in a chilling phrase. "Religion is poison," he told His Holiness. "It neglects material progress."

Following the return of the Dalai Lama to Lhasa, an uneasy calm enveloped the ancient capital. Then, in 1959, the possibility of coexistence between Tibet and China was shattered in a morning as word of an uprising in the eastern provinces of Kham and Amdo sparked open rebellion in the streets of Lhasa. Disguised as a peasant, the Dalai Lama rode away from the city on the first leg of an epic journey that would, in time, carry him across the Himalayan snows into exile in India. In his wake, he left a trail of sorrow soon to be followed by tens of thousands of his countrymen.

With the Tibetan leadership broken and the provisions of the 17 Point Agreement nullified by events, all talk of respecting traditions and honoring local political autonomy ceased. The Chinese were free to pursue their ultimate goal, which was nothing less than the total revolution of Tibetan society.

The Anti-Rebellion campaign that ensued both quelled the revolt and spread fear and instability throughout the land. In scores of villages that had yet to encounter the invaders, heavily armed Chinese cadres arrived to institute *thumzing*, struggle sessions in which the old ways were denigrated and the communists promoted as liberators. The intent was the ideological conversion of the people and the subversion of all traditional institutions, encapsulated in "the three evils": the aristocratic land owners, the Dalai Lama's government, and the monasteries, then home to one Tibetan out of ten.

Public meetings degenerated into theatrical whirlwinds of false accusations and the settling of old scores. Monks and diviners, oracles and high lamas, were paraded in dunce hats, the object of cruel and ruthless ridicule. Under the watchful eyes of the Chinese army, Tibetan farmers and nomads learned to recite Maoist slogans as effortlessly as they had once chanted Buddhist prayers. Even in lands where no one had risen in revolt, monks and landowners were condemned for high treason. The entire Tibetan army was marched off to labor camps. A people who for centuries had believed in the holy character of their monasteries were told that chörtens and temples were but prisons of mud and dirt, the esoteric and sacred knowledge of the priests but tangled words of entrapment. Ideologically committed to the eradication of religion, the Chinese vigorously attacked the monastic order, forcibly disbanding the large estates and undercutting at every opportunity the status and authority of the more than a hundred thousand monks and nuns. The destruction of the economic basis of the monasteries would prove to be the most significant social and political event in Tibetan history since the introduction of Buddhism.

Ideological fanaticism, materialist thought control, and communist class struggle reached a watershed during the Cultural Revolution, a cataclysmic upheaval that unleashed a wave of barbarism that in the end even Mao himself could not control. The campaign began in 1966 as a purge of the Communist Party leadership, a deliberate infusion of instability designed to perpetuate revolutionary zeal and secure Mao's place as the helmsman and sole mediator of the nation. Its ostensible goal was the creation of a pure socialist cadre, men and women whose minds had been purged and memories erased to yield a template upon which the thoughts of Mao could be engraved. The true and just society would emerge in the wake of the destruction of the Four Olds: old ideas, old culture, old customs, old habits. Create the new by smashing the old. This was the official slogan of what was heralded to be the last battle before the coming of the socialist paradise.

All notions of religion and spirit, the poetics of culture and family, intuitions about the relationship of man and woman and nature, the scent of the soil, and the meaning of rain falling upon stones had no place in Mao's calculus of transformation and domination. Nationality was considered a mere product of economic disparity; once material inequalities had been addressed, ethnic distinctions would wither. Tibet, of course, exemplified the old; China, the new. Thus, the Cultural Revolution both implied and demanded a total assault on every facet of Tibet's ancient civilization.

In villages throughout the land, people were mobilized to destroy chörtens and temples, prayer flags, incense burners, religious statuary, prayer wheels, and *mani* stones. Sacred objects centuries old were shattered, thousands of monasteries vandalized. At Lithang, six thousand Tibetans locked themselves inside the monastery gates and endured a siege of sixty-four days, until finally a massive Chinese airstrike left four thousand dead and the structure in ruins. At Ganden, a place of spiritual intensity and architectural grandeur second only to the Potola in Lhasa, Chinese cadres used dynamite and artillery fire to reduce to rubble the entire complex, a city of learning at the time home to four thousand clergy. In multiple acts of blasphemy, thousands of scriptures were set ablaze, and those not burned were dispensed to shops to be used as wrapping paper or as the liners and padding for shoes. In Lhasa, sacred texts littered the ground around public latrines, where the Chinese left them in bundles to be used as toilet paper.

By decree, the recitation of prayers ceased. The practice of prostration and the circumambulation of holy sites was outlawed, as was the refined language of the aristocracy. Those who sought treatment from traditional healers, or knowledge of the future through consultation with oracles and diviners, faced severe sanctions. Gift giving, the ceremonial exchange of silk khata, symbol of hope and goodwill, was prohibited. Even the picnics and parties for which Tibetans were famous were forbidden as symbols of the feudal past. Ideological fervor challenged even the traditional Tibetan way of using a shovel, whereby one person worked the tool and a second helped lift the load by means of a rope attached to the spade. It had been legitimate, the cadre suggested, to save energy whilst working for the landlords, but in the new China, everyone had to work with full vigor for the motherland and the socialist state. Thus, the ropes had to go.

In a frenzy of fear, Tibetans struggled to prove their revolutionary zeal. Streets and parks in Lhasa sprouted new names, as did children. The elegant clothing of old Tibet was replaced with the drab attire of Mao, uniforms of gray and green cloth upon which people deliberately sewed unnecessary patches to affirm their proletarian credentials. *Om mani padme hum*, the "jewel in the lotus," the mantra of Buddhist compassion carved into stone and recited in temples and caves for a thousand years, was supplanted by the slogan of a personality cult: *Mao Zedong wan sui*, "May Mao live for ten thousand years." For those who resisted, there was sadistic punishment: people burned alive, people forced to eat human waste, tortures too numerous and dreadful to enumerate.

By 1970, the process of transformation appeared complete. The monasteries and temples, as many as six thousand sacred monuments and centers of wisdom and veneration, lay in ruins. The monks who had once dominated the nation were gone. Overt religious practices of any sort had ceased. The divisions between rich and poor, educated and ignorant, powerful and weak, had been dissolved by decree. Endless rounds of political meetings intimidated the elders. Children were separated from their families to attend state schools, where revolutionary zeal was the measure of pedagogy. Superficially, at least, Tibetans ceased to have any distinctive characteristics. All diversity of spirit and heart, of thoughts and imaginings, dissolved into shadows as a cloak of conformity fell upon the land.

The cruel excesses of the Cultural Revolution afflicted not only Tibet, of course, but all of China. So widespread were the injustices and so numerous the tormentors that, in order to survive, modern China has simply expunged the era from its memory, a bitter legacy swept beneath the carpet of history. In Tibet, such sleight of hand is not possible. Over a million Tibetans were killed. There are simply too many memories among the families of the dead, too many signs of physical destruction upon the land: the broken stones and walls of Shegar Dzong, the Shining Crystal monastery that once held the finest collection of woodcut books in Tibet, or the landslide of rubble on the immense flank of the Drokpa Ri ridge, all that remains of the original Ganden monastery, a warren of hundreds of structures built over the course of half a millennium and destroyed in a month.

The suggestion that such devastation was the price of modernity and that Tibet at least escaped the tyranny of feudalism and benefited by the economic development brought by the Chinese deserves consideration. It is certainly true that many Tibetans loathed the inequities and exploitation of the monastic era. I once spent a month on the eastern approaches to Everest with a Tibetan who had fled as a youth to India for economic reasons. With anger and contempt, he recalled watching his relatives carrying petrol and dismantled sections of vehicles on their backs as they embarked on a journey of hundreds of miles so that the Dalai Lama and his entourage could play with their cars in the streets of Lhasa. He remembered the prisoners let out at night from the local jail to beg for their evening meal, men in wooden yokes, many with severed kneecaps and stumps for hands. Electricity and running water were unknown. When he and his family reached Darjeeling, he had played for hours with a water faucet and stared in delight at a light bulb. In Tibet, he recalled, the ruling elite had offered them nothing but prayer wheels. What they had wanted were real wheels.

Roads the Chinese did bring to Tibet, twelve hundred miles by 1954 alone. These arteries allowed for the garrisoning of troops, the movement into Tibet of Chinese settlers, and the shipment of timber and other resources from the mountains to the densely populated provinces of western China. But they also facilitated the flow of goods and people within Tibet itself. The Chinese improved primary health care, introduced electricity to remote villages, and provided land to previously disenfranchised peasants, all initiatives that were well received by the majority of Tibetans.

But a closer examination of the consequences of the Chinese occupation reveals that their efforts concentrated on ideological education and social transformation rather than industrial and economic development. Political chaos destroyed the traditional economic system, leaving little in its place. The forced collectivization of land proved in the end to be a disaster. An edict to replace barley with wheat led only to hunger, leaving the predominantly agricultural nation in abject poverty. Even the fundamental need to retain some grain reserves in anticipation of unexpectedly harsh winters and the need to sow another crop in the spring was lost in the fervor of revolution. Far from eradicating poverty, the communists left the nation in ruins through their economic mismanagement. Not until 1981, five years after the death of Mao, did economic reforms finally return the Tibetan people to the standard of living they had enjoyed more than thirty years before the arrival of the Chinese.

The notion that Tibetan society would have remained static during those years and that the material lot of the people would not have improved without the Chinese is highly questionable. The monastic orders were deeply conservative, but not all elements of the Tibetan elite were resistant to innovation. The Dalai Lama's own family had begun to disband its estates and redistribute land long before the Chinese invasion. His Holiness himself was famously intrigued by Western science and technology, and with the winds of change blowing harshly upon his borders throughout the middle years of this century, it is most unlikely that the Tibetans would have either wanted or been able to resist the intrusion of modernity.

The Chinese investment over the years has indeed been monumental. By 1994, more than $4 billion had been spent, $600 million in 1996 alone. Just what has been achieved and what has been lost remains in dispute. Between 1982 and 1991, literacy rates actually declined under Chinese rule. Much of the financial aid has either been squandered on ill-conceived projects or been used to facilitate the transformation of a civilization. As a result, Lhasa has become a Chinese enclave, with the economy dominated by Han immigrants.

In a brazen effort to tear from the fabric of the ancient capital its resonance and identity, Chinese authorities endorsed a campaign of modernization that by 1996 had resulted in the destruction of over half of the historic buildings in the city. In their place rose stark Chinese structures of concrete and chrome, all blue mirrors and cheap tiles, with corridors where the cement dust never settles. In the Barkhor, the market bazaar that surrounds the Jokhang, most sacred of all Tibetan temples, Asian pop music today drowns out the prayers of pilgrims. In doorways, Chinese merchants crouch in the dirt, drinking long draughts of cold tea, spitting and coughing as they snap their cigarette ashes into the paths of monks.

In the history of modern Tibet, the brutality and suffering are overwhelming, but in a certain sense what is equally disturbing is the realization that the Maoists actually believed in what they were doing. Rhetoric that today appears laughable was embraced as gospel. Their cause was from the start a totalitarian fantasy. The goal was the triumph of materialism, a deterministic model of the world that left no place for the finer sensibilities of the human spirit.

What is truly remarkable in this long and tragic saga is the resilience of the Tibetan people, the redemptive power of their faith, and the strength of their determination to survive as a nation. The reforms initiated by the Chinese in the early 1980s revealed that all the years of torment and propaganda had done little to fundamentally transform the culture. Once free of the overt repression of the Maoist era, the Tibetans within weeks discarded the veneer of proletarian life imposed by the communists. In every village, religious relics reappeared, family treasures surfaced from the soil, and men and women began the stone-by-stone task of rebuilding the chörtens and temples. The Chinese spoke of duty to the motherland, but the Tibetans clearly felt no filial piety. What they honor is the Buddha. Their strength grows out of a conviction that suffering can only be overcome by acceptance of the impermanence and illusionary nature of reality. In the Diamond Sutra, the Buddha cautions that the world is fleeting, like a candle in the wind, a phantom, a dream, the light of stars fading with the dawn. It is upon this insight that Tibetans measure their past and chart their future.

A yak herder gathering fodder,
Kama Valley, Tibet, 1997

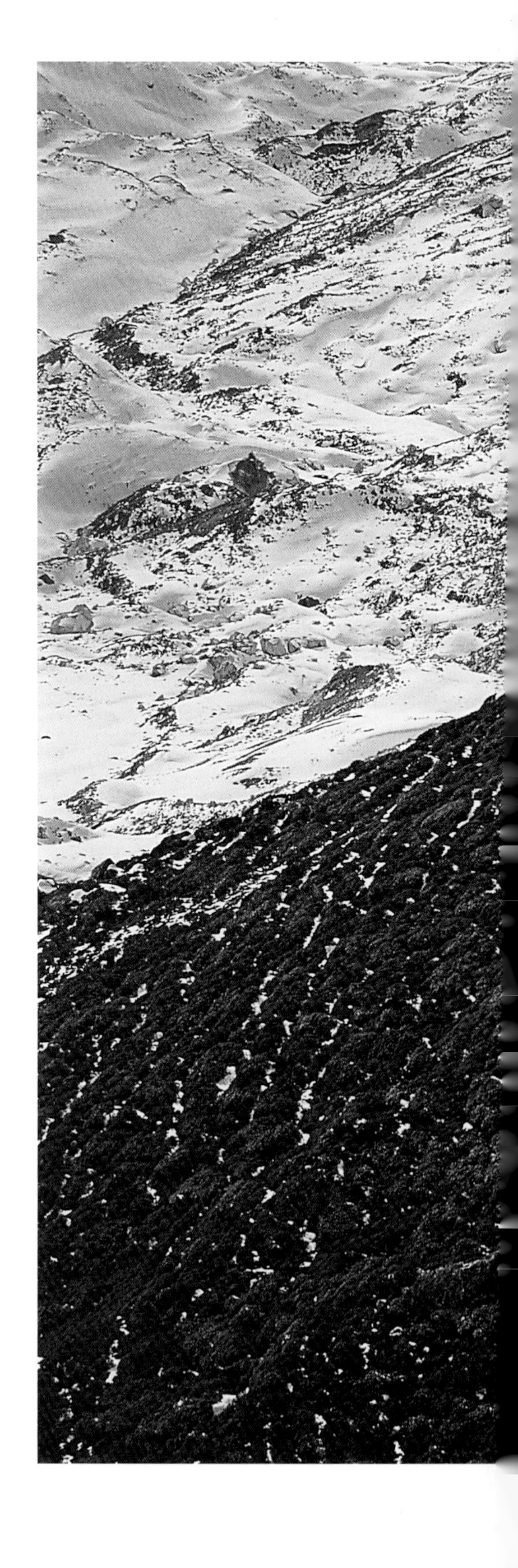

A yak train approaches the Kangshung face of Chomolungma, Kama Valley, Tibet, 1997

A man and his grandson painting the walls of the
Tashilhunpo Monastery, Shigatse, Tibet, 1997

A prayer wheel and a temple guardian,
Shekar, Tibet, 1997

Monks chanting prayers for the Buddha's birthday,
Boudhnath, Kathmandu, 2000

Buddhist nuns prepare offerings at Jokhang Temple,
Lhasa, Tibet, 1996

Pilgrims at the Pango Chörten, Gyantse, Tibet, 1996

A young Buddhist monk greets the dawn,
Shekar Monastery, Tibet, 1997

Monks at Angkor Wat, Cambodia, 2001

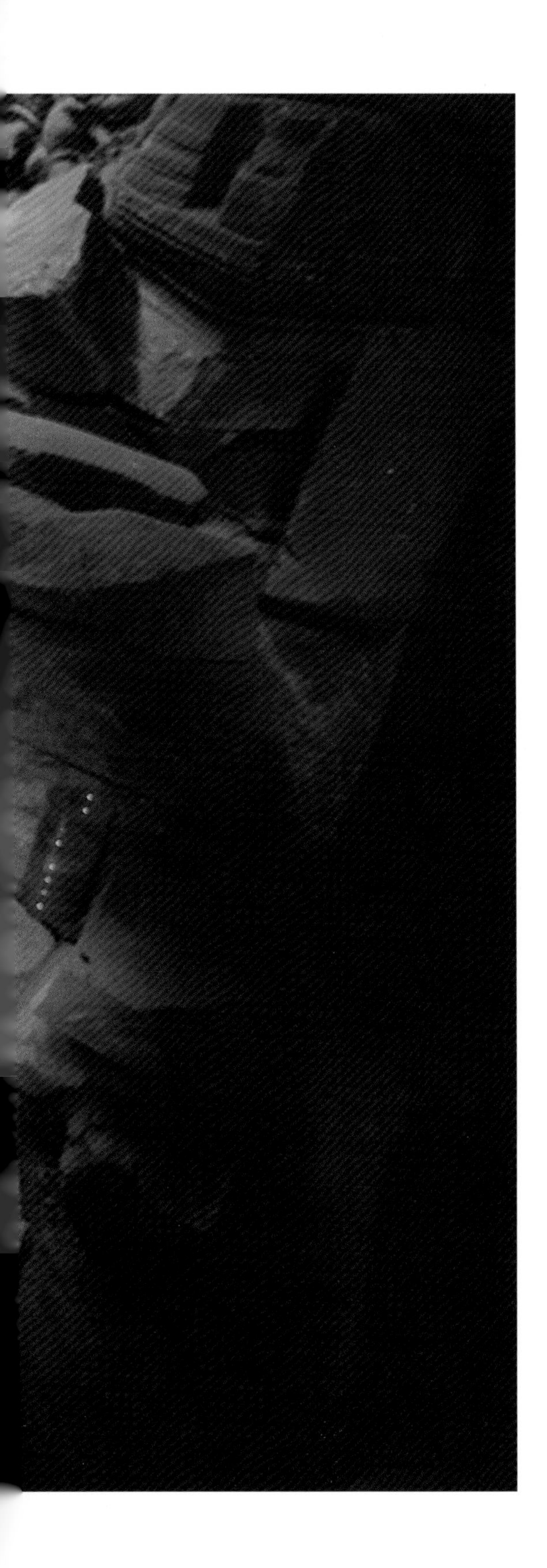

A temple guardian, Ta Prohm, Cambodia, 2001

A young Caboclo girl in the delta of
the Amazon, Brazil, 1987

CHAPTER SEVEN

A THOUSAND WAYS OF BEING

Like the Chinese in Tibet, the Europeans who colonized Australia were unprepared for the sophistication of the place and its inhabitants, incapable of embracing its wonder. They had no understanding of the challenges of the desert and little sensitivity to the achievements of aboriginal peoples who for over fifty thousand years had thrived as nomads, wanderers on a pristine continent. In all that time, the desire to improve upon the natural world, to tame the rhythm of the wild, had never touched them. The Aborigines accepted life as it was, a cosmological whole, the unchanging creation of the first dawn, when earth and sky separated, and the original Ancestor brought into being all the primordial Ancestors who, through their thoughts, dreams and journeys, sang the world into existence.

The Ancestors walked as they sang, and when it was time to stop, they slept. In their dreams, they conceived the events of the following day, points of creation that fused one into another until every creature, every stream and stone, all time and space, became part of the whole, the divine manifestation of the one great seminal impulse. When they grew exhausted from their labors, they retired into the earth, sky, clouds, rivers, lakes, plants, and animals of an island continent that resonates with their memory. The paths taken by the Ancestors have never been forgotten. They are the Songlines, precise itineraries followed even today as the people travel across the template of the physical world.

As the Aborigines track the Songlines and chant the stories of the first dawning, they become part of the Ancestors and enter the Dreamtime, which is neither a dream nor a measure of the passage of time. It is the very realm of the Ancestors, a parallel universe where the ordinary laws of time, space, and motion do not apply, where past, future, and present merge into one. It is a place that Europeans can only approximate in sleep, and thus it became known to the early English settlers as the Dreaming, or Dreamtime. But the term is misleading. A dream by Western definition is a state of consciousness divorced from the real world. The Dreamtime, by contrast, is the real world, or at least one of two realities experienced in the daily lives of the Aborigines.

To walk the Songlines is to become part of the ongoing creation of the world, a place that both exists and is still being formed. Thus, the Aborigines are not merely attached to the Earth, they are essential to its existence. Without the land, they would die. But without the people, the ongoing process of creation would cease and the Earth would wither. Through movement and sacred rituals, the people maintain access to the Dreamtime and play a dynamic and ongoing role in the world of the Ancestors.

A moment begins with nothing. A man or a woman walks, and from emptiness emerge the songs, the musical embodiment of reality, the cosmic melodies that give the world its character. The songs create vibrations that take shape. Dancing brings definition to the forms, and objects of the phenomenological realm appear: trees, rocks, streams, all of them physical evidence of the Dreaming. Should the rituals stop, the voices fall silent, all would be lost. For everything on Earth is held together by the Songlines, everything is subordinate to the Dreaming, which is constant but ever changing. Every landmark is wedded to a memory of its origins and yet always being born. Every animal and object resonates with the pulse of an ancient event, while still being dreamed into being. The world as it exists is perfect, though constantly in the process of being formed. The land is encoded with everything that ever has been, everything that ever will be, in every dimension of reality. To walk the land is to engage in a constant act of affirmation, an endless dance of creation.

The Europeans who first washed ashore on the beaches of Australia lacked the language or imagination even to begin to understand the profound intellectual and spiritual achievements of the Aborigines. What they saw was a people who lived simply, whose technological achievements were modest, whose faces looked strange, whose habits were incomprehensible. The Aborigines lacked all the hallmarks of European civilization. They had no metal tools, knew nothing of writing, had never succumbed to the cult of the seed. Without agriculture or animal husbandry, they generated no surpluses, and thus had never embraced sedentary village life. Hierarchy and specialization were unknown. Their small semi-nomadic bands, living in temporary shelters made of sticks and grass, dependent on stone weapons, epitomized European notions of backwardness. An early French explorer described them as "the most miserable people of the world, human beings who approach closest to brute beasts." As late as 1902, a member of the Australian parliament claimed, "There is no scientific evidence that the Aborigine is a human at all."

By the 1930s, a combination of disease, exploitation, and murder had reduced the Aborigine population from well over a million at the time of European contact to a mere thirty thousand. In one century, a land bound together by Songlines, where the people moved effortlessly from one dimension to the next, from the future to the past and from the past to the present, was transformed from Eden to Armageddon.

Knowing what we do today of the extraordinary reach of the Aboriginal mind, the subtlety of their thoughts, and the evocative power of their rituals, it is chilling to think of this reservoir of human potential, wisdom, intuition, and insight that very nearly ran dry during those terrible years of death and conflagration. As it is, Aboriginal languages, which may have numbered 250 at the time of contact, are disappearing at the rate of one or more per year. Only eighteen are today spoken by as many as five hundred individuals.

Despite this history, the Aborigines have survived and, in time, may still have a chance to inspire and redeem a nation. Reconciliation and the building of partnerships between Aborigines and non-Aborigines, including serious efforts to resolve land disputes and address historical wrongs, today dominate the Australian political agenda. The languages and cultures of the Aborigines are taught in universities, extolled in popular song, featured in films.

Aboriginal notions of design and decoration, transferred to canvas in the 1970s, have given rise to an art form celebrated throughout the world. In a symbolic moment, charged with emotion, an Aboriginal athlete inspired the world when selected to ignite the ceremonial flame at the Sydney 2000 Olympic Games.

~

BUT WHAT OF THE OTHER VICTIMS of European expansion? The people of Tasmania were exterminated within seventy-five years of contact. The Reverend John West, a Christian missionary, rationalized their slaughter: "Their appearance is offensive, their proximity obstructive, their presence renders everything insecure. Thus the muskets of the soldier, and those of the bandits, are equally useful; they clear the land of a detested incubus."

Within a generation of Captain James Cook's landing in Oahu, only thirty thousand Hawaiians survived out of an original population of some eight hundred thousand. In the Caribbean, on the island of Hispaniola, the Arawakan population of well over a million was eliminated within fifteen years.

To take full measure of what contact implied for the peoples of the Americas, recall that the word "decimate," horrific as it is intended to be, means to kill one in ten. Within three generations of contact, over 90 percent of Native Americans, people of hundreds of nations, each with its own vision of reality, living from the Arctic to Tierra del Fuego, had succumbed to measles, smallpox, and other European diseases that spread like an evil miasma. In central Mexico, the population, densest in the Americas, collapsed from twenty-five million to two million within sixty years of the conquest of the Aztec. As late as 1800, California had an indigenous population of three hundred thousand; by 1950, their descendants numbered only ten thousand. Nine million died in Peru. Twenty-three million in Mexico. Another five million in the Amazon alone.

Echoes of this holocaust are still heard in every corner of the globe. Indeed, in many lands the carnage continues. Throughout this century, indigenous societies in Brazil have disappeared at the rate of one per year. In Colombia, as recently as 1974, a group of white miners appeared in court charged with the murder of several Cuiva Indians. In their defense, they professed not to believe that the Cuiva were humans.

In the upper reaches of the Orinoco, a gold rush brings disease to the Yanomami, killing a quarter of the population in a decade, leaving the survivors hungry and destitute. In Nigeria, pollutants from the oil industry so saturate the floodplain of the Niger River, homeland of the Ogoni, that the once fertile soils can no longer be farmed. A vortex of violence and famine in the Sudan claims the lives of tens of thousands of Nuer. The Efe pygmies, forest dwellers of the Congo, dwindle toward extinction as sexually transmitted diseases ravage the population. In Siberia, Soviet authorities, frustrated by their inability to control the nomadic Ul'ta, oblige the children of the reindeer herders to attend Russian language schools, thus severing the bonds of family and violating a way of life. Today, only a single herding group remains, ten Ul'ta men who still follow ancestral migratory routes so familiar to their animals that the reindeer lead the way.

These are not isolated events but elements of a global phenomenon that will no doubt be long remembered as one of the tragedies of this age. As essential to humanity as the environments they steward, indigenous societies throughout the world are under siege. Well over half are moribund.

The notion that these societies are simply fated to fade away is quite wrong. In virtually every instance, these indigenous peoples are being torn from their past and propelled into an uncertain future because of specific political and economic decisions made by powerful outside entities.

The plight of the Penan in Sarawak, as we have seen, is the result of government policies that promote and reward the unsustainable extraction of timber from a forest that was once their homeland. The case of the Rendille in Kenya reads as a parable for what can happen when, in the name of development, good intentions come together with bad ideas. Tibet is a tale of brute conquest. That all of these conflicts result from deliberate choices made by men and women is both discouraging and empowering. If people are the agents of destruction, they can also be the agents of cultural survival. Happily, there are examples of redemption that can inspire us all.

IN THE ARCTIC, one marvels at the art of survival. Bears hunt seals, foxes follow the bears and feed on their excrement. Inuit women cut open animals to feed on clam siphons found in walrus stomachs, lichens and plants in the guts of caribou, mother's milk in the bellies of baby seals. They store meat taken in August in skins and bladders cached in rock cairns, where it ferments to the consistency and taste of blue cheese.

In winter darkness, when temperatures fall so low that breath cracks in the wind, Inuit men leave their families to follow the open leads in the new ice; there, they kneel motionless for hours at a time over the breathing holes of ringed seals. The slightest shift in weight will betray their presence, so they squat in perfect stillness, all the while knowing full well that as they hunt, they are being hunted. Polar bear tracks run away from every breathing hole. If a seal does not appear, the hunter may roll over, mimicking a seal to try to attract a bear so that predator may become prey.

The northern landscape is empty and desolate. On the horizon, islands, ice, and sky meld one into the other, and the black sea is a distant mirage. But the Inuit seldom lose their way. In driving snowstorms, they watch for patterns in the ice, small ridges of hard snow that run parallel to the prevailing winds and reveal where they are. They study a map of the land reflected on the underside of low clouds: open water is black; the sea ice, white; ground covered in snow and traces of open tundra are darker than the sea but lighter than snowless land.

When the Inuit first encountered Europeans, they mistook their wooden ships with billowing sails for gods. The white men viewed the natives as savages. Both were wrong, though the Inuit did more to dignify the human race. The entire history of European exploration was colored by a single theme: those who ignored the example of the Inuit perished, whereas those who imitated their ways not only survived but achieved great feats of endurance and discovery. In the end, it was the English who suffered most for their arrogance, dying by the score at the entrance to the Northwest Passage.

When the last of John Franklin's men died at Starvation Cove on the Adelaide Peninsula, their sledge alone weighed 650 pounds. On it was an 800-pound boat loaded with silver dinner plates, cigar cases, a copy of *The Vicar of Wakefield*—in short, necessities for a gentleman traveler of the Victorian age. In the leather traces of the sledge were the frozen corpses of young British sailors. Scorning the use of dogs, the English had planned to haul this cumbersome load hundreds of miles overland in the hope of reaching some remote trading post in the vast boreal forests of Canada. Like so many of their kind, as one explorer remarked, they died because they brought their environment with them and were unwilling to adapt to another.

In dismissing the Inuit as savages, the British failed to grasp the measure of intelligence necessary to thrive in the Arctic with a technology limited to what could be made with ivory and bone, antler and animal skins, soapstone and slate, and precious bits of driftwood. The Inuit did not merely endure the cold, they took advantage of it. Three Arctic char placed end to end, wrapped and frozen in hide, the bottom greased with the stomach contents of a caribou and coated with a thin film of ice, became the runner of a sled. A knife could be made from human excrement. There is a well-known account of an old man who refused to move into a settlement. Fearful for his life and hoping to force him off the ice, his family took away all of his tools and weapons. So, in the midst of a winter gale, he stepped out of their igloo, defecated, and shaped the feces into a frozen blade, which he sharpened with a spray of saliva. With the knife, he killed a dog. Using its ribcage as a sled and its hide to harness another dog, he disappeared into the darkness.

Blind to the genius of the Inuit, the early European arrivals imposed their own order. Along with traders who brought diseases that killed nine of ten Inuit, came missionaries whose primary goal was to destroy the power and authority of the shaman, the cultural pivot, the symbolic heart of the Inuit relationship to the universe. The priests forbade the use even of traditional names, songs, and the language itself.

By the early twentieth century, the seduction of modern trade goods had begun to draw people away from the land. As they concentrated in communities, often around missions and trading posts, new problems arose. A distemper outbreak led Canadian authorities to rationalize the wholesale slaughter of sled dogs, in whose place the Royal Canadian Mounted Police introduced snowmobiles, with the first one arriving on Baffin Island in 1962. In the settlements, health conditions deteriorated. A tuberculosis outbreak in the 1950s, and the desperate attempts by medical authorities to curtail its spread, resulted in 20 percent of the entire Inuit population being forcibly removed to sanitaria in Montreal, Winnipeg, and other southern cities.

In two decades of destruction, beginning in the late 1940s, the Inuit drifted toward dependency. As the Cold War escalated, the Canadian government, compelled to bolster its claims to the Arctic, actively promoted settlement. Family allowance payments were made contingent on the children attending school. As parents moved into communities to be with their young, nomadic camps disappeared. Along with settlements and schools came nursing stations, churches, and welfare. The government identified each person by number, issued identification tags, and ultimately conducted Operation Surname, assigning last names to a people who had never had them. In the process, more than a few Inuit dogs were recorded as Canadian citizens.

After half a century of profound changes, what has become of Inuit traditions? Naturally, the people have adapted. The Inuit language is alive. The men are still hunters: they use snares, make snow houses, know the power of medicinal herbs that sprout in the Arctic spring. But they also own boats, snowmobiles, television sets, and satellite phones. Some drink, some attend church. As anthropologist Hugh Brody points out, what must defended is not the traditional as opposed to the modern, but, rather, the right of a free indigenous people to choose the components of their lives.

Canada, which had not always been kind to the Inuit, has in recent years recognized this challenge by negotiating an astonishing land-claims settlement with the peoples of the Eastern Arctic. In a historic gesture of restitution, the federal government announced on April 1, 1999, the creation of Nunavut, an Inuit homeland of well over 770,000 square miles. Including all of Baffin Island and stretching from Manitoba to Ellesmere Island, with a population of just twenty-six thousand, Nunavut is almost as large as Alaska and California combined. In addition to annual payments totaling $1.148 billion over fourteen years, funds to be held in trust to finance student scholarships and economic development, the Inuit will receive direct title to over 135,000 square miles. More than 80 percent of the known mineral reserves of copper, lead, zinc, gold, and silver in all

On the ice of Lancaster Sound at Cape Crauford,
Baffin Island, Nunavut, 1997

Olayuk and Martha Narqitarvik at Cape Crauford,
Baffin Island, Nunavut, 1997

The midnight sun at Cape Crauford, Baffin Island, Nunavut, 1997

of Nunavut are to be found on land deeded to the Inuit. In their homeland, where caribou outnumber people thirty to one and where but a century ago nomadic hunters fashioned tools from stone and slate, the Inuit will have total administrative control, the most remarkable experiment in Native self-government anywhere to be found.

There is much to do, many injuries to heal. After years of decay and alienation, substance abuse is chronic and the suicide rate is six times the national average. When the environmental group Greenpeace shut down the traditional seal hunt on Baffin Island, the annual per capita income dropped from $16,000 to nothing. Unemployment in the cash economy of Nunavut hovers at 30 percent.

But if the challenges are great, so is the opportunity. In the language of the Inuit, the word *uvatiarru* may be translated as "long ago" or "in the future." A cultural renaissance is underway, with the Inuit at last in control of their lives and destiny. Tracking caribou on the open tundra during the cold months of the fall, taking narwhal from the ice in July, they will continue to honor a seasonal round that recalls a not too distant era when their people were nomads, hunters of the northern ice. The rhythm of the year will continue to propel the culture, influencing everything from the character of local regulations and the criminal justice system to the manner in which men and women come together, families grow and prosper, power is shared and food divided. In time, the wounds of broken promises and lost dreams, the tortured memories of the residential schools, where children were torn from families and young boys and girls lived at the mercy of priests, will be healed. It may take generations. But then patience has always been one of the most enduring traits of the Inuit. There is a story from Greenland about a group of men and women who walked a great distance to gather wild grass in one of the few verdant valleys of the island. When they arrived, the grass had yet to sprout, so they watched and waited until it grew.

~

As you read these last words, and perhaps glance over the photographs once again, consider that everything that has been expressed in this book, every story and myth, each point of conflict and tragedy, has been distilled from my own limited experiences and from what anthropologists have learned from their work among the handful of cultures that it has been my privilege to have known. People sometimes remark that I have been fortunate to have traveled so widely, to have seen so much, but, in fact, I have experienced very little of the world's cultural diversity. Just as a bouquet of a dozen species represents but an iota of the Earth's botanical bounty, so too these journeys and encounters offer but a fragment of the full wonder of the ethnosphere. In this book I make reference to perhaps thirty cultures, discussing at some length a mere fourteen. There remain another fourteen thousand to visit and celebrate, if only there were time.

Before she died, anthropologist Margaret Mead spoke of her singular concern that, as we drift toward a more homogenous world, we are laying the foundations of a blandly amorphous and singularly generic modern culture that ultimately will have no rivals. The entire imagination of humanity, she feared, might become confined within the limits of a single intellectual and spiritual modality. Her nightmare was the possibility that we might wake up one day and not even remember what had been lost.

There are many people, of course, who view such a process of condensation as progress, the inevitable consequence of modernity. Without doubt, images of comfort and wealth, of technological wizardry, have a magnetic allure. Any job in the city may seem better than backbreaking labor in sun-scorched fields. Entranced by the promise of the new, indigenous people throughout the world have in many instances voluntarily and in great earnest turned their backs on the old. The consequences can be profoundly disappointing. The fate of the vast majority of those who sever their ties with their traditions will not be to attain the prosperity of the West but to join the legions of urban poor, living on the edges of a cash economy, trapped in squalor, and struggling to survive. As cultures wither away, individuals remain, often shadows of their former selves, caught in time, unable to return to the past yet denied any real possibility of securing a place in the world whose values they seek to emulate and whose wealth they long to acquire.

The triumph of secular materialism is the conceit of modernity. But what are the features of this life? An anthropologist from a distant land visiting America, for example, would note many wondrous things. But he would no doubt be puzzled to learn that 20 percent of the people control 80 percent of the wealth, that the average child has by the age of eighteen spent a full two years passively watching television. Observing that over half of our marriages end in divorce and that only 6 percent of our elders live with a relative, he might question the values of a society that so readily breaks the bonds of marriage and abandons its aged, even as its men and women exhaust themselves in jobs that only reinforce their isolation from their families. Certainly, a slang term such as 24/7, implying as it does the willingness of an employee to be available for work at all times, seems excessive, though it would explain the fact that the average American father spends only eighteen minutes a day in direct communication with his child. And what of our propensity to compromise the very life supports of the planet? Extreme would be one word for a civilization that contaminates with its waste the air, water, and soil; that drives plants and animals to extinction; that dams the rivers, tears down the ancient forests, rips holes in the protective halo of the heavens, and does little to curtail industrial processes that threaten to transform the biochemistry of the very atmosphere.

Once we look through the anthropological lens and see, perhaps for the first time, that all cultures have unique attributes that reflect choices made over generations, it becomes absolutely clear that there is no universal progression in the lives and destiny of human beings. No trajectory of progress. Were societies to be ranked on the basis of technological prowess, the Western scientific experiment, radiant and brilliant, would no doubt come out on top. But if the criteria of excellence shifted, for example, to the capacity to thrive in a truly sustainable manner, with a true reverence and appreciation for the Earth, the Western paradigm would fail. If the imperatives driving the highest aspirations of our species were to be the power of faith, the reach of spiritual intuition, the philosophical generosity to recognize the varieties of religious longing, then our dogmatic conclusions would again be found wanting.

Viewed from this broader perspective, the notion that indigenous societies are archaic, that their very presence represents some impediment, is transparently wrong. As David Maybury-Lewis has written, indigenous peoples do not stand in the way of progress; rather, they contribute to it if given a chance. Their cultural survival does not undermine the nation state; it serves to enrich it, if the state is willing to embrace diversity. And, most important of all, these cultures do not represent failed attempts at modernity, marginal peoples who somehow missed the technological train to the future. On the contrary, these peoples, with their dreams and prayers, their myths and memories, teach us that there are indeed other ways of being, alternative visions of life, birth, death, and creation itself. When asked the meaning of being human, they respond with ten thousand different voices. It is within this diversity of knowledge and practice, of intuition and interpretation, of promise and hope, that we will all rediscover the enchantment of being what we are, a conscious species aware of our place on the planet and fully capable not only of doing no harm but of ensuring that all creatures in every garden find a way to flourish.

A Hindu saddhu, Bhaktapur, Nepal, 2001

A Buddhist monk in Chamdo, Tibet, 1996

The healing waters of Damballah,
Saut D'eau, Haiti, 1982

ACKNOWLEDGEMENTS

IN MY TRAVELS, I was never alone, and throughout all of these adventures and investigations, I benefited enormously from the companionship and guidance of dozens of individuals. I am especially indebted to the men and women of the indigenous communities who have so generously and openly received me as a guest in their lives.

While it would be impossible to list by name all those who have assisted me over these many years, I would like to express my gratitude to my teachers and mentors, friends and colleagues. In particular, I would like to pay tribute to four people who have died recently: Darell Posey, tireless champion of the Kayapó and the human rights of all Amazonian peoples; Bruno Manser, friend and defender of the Penan; Terence McKenna, mystic visionary; and Richard Evans Schultes, plant explorer and teacher.

I am also grateful to the photographers who have taught and inspired me, and to the editors and designers responsible for working with the words and images to create this book. Finally, I thank my family: Gail, Tara, and Raina.

I would like to dedicate this book to my sister, Karen, for her unfailingly generous love and support.

Light at the Edge of the World
Wade Davis

Published by the National Geographic Society
John M. Fahey, *President and Executive Officer*
Gilbert M. Grosvenor, *Chairman of the Board*
Nina D. Hoffman, *Executive Vice President*

Text and photographs copyright © 2001 Wade Davis. All rights reserved.
Reproduction of the whole or any part of the contents without permission is prohibited.

Originated and published in Canada by Douglas & McIntyre,
2323 Quebec Street, Suite 201, Vancouver, British Columbia V5T 4S7

Editing by Saeko Usukawa
Design by Val Speidel and Marianne Koszorus
Jacket photographs by Wade Davis
Printed and bound in Canada by Friesens
Printed on acid-free paper ∞

The world's largest nonprofit scientific and educational organization, the National Geographic Society was founded in 1888 "for the increase and diffusion of geographic knowledge." Since then it has supported scientific exploration and spread information to its more than eight million members worldwide.

The National Geographic Society educates and inspires millions every day through magazines, books, television programs, videos, maps and atlases, research grants, the National Geographic Bee, teacher workshops, and innovative classroom materials.

The Society is supported through membership dues, charitable gifts, and income from the sale of its educational products.

Members receive NATIONAL GEOGRAPHIC magazine—the Society's official journal—discounts on Society products, and other benefits.

For more information about the National Geographic Society, its educational programs, publications, or ways to support its work, please call 1-800-NGS-LINE (647-5463), or write to the following address:

National Geographic Society
1145 17th Street N.W.
Washington, D.C. 20036-4688 U.S.A.

Visit the Society's Web site at www.nationalgeographic.com.

10

306.08
D28
Davis, Wade.
Light at the edge of the world: a journey through the realm of vanishing cultures.
c.1

CHICAGO HEIGHTS FREE PUBLIC LIBRARY
25 WEST 15TH ST.
CHICAGO HEIGHTS, ILL.
60411
PHONE: (708) 754-0323

GAYLORD S